Basic Statistical Methods and Models for the Sciences

Judah Rosenblatt

CRC Press
Taylor & Francis Group
Boca Raton London New York

CRC Press is an imprint of the
Taylor & Francis Group, an **informa** business

A CHAPMAN & HALL BOOK

Basic Statistical
Methods
and Models
for the Sciences

CRC Press
Taylor & Francis Group
6000 Broken Sound Parkway NW, Suite 300
Boca Raton, FL 33487-2742

First issued in paperback 2019

© 2002 by Taylor & Francis Group, LLC
CRC Press is an imprint of Taylor & Francis Group, an Informa business

No claim to original U.S. Government works

ISBN-13: 978-1-58488-147-6 (hbk)
ISBN-13: 978-0-367-39645-9 (pbk)

Library of Congress Cataloging-in-Publication Data

Catalog record is available from the Library of Congress

Visit the Taylor & Francis Web site at
http://www.taylorandfrancis.com

and the CRC Press Web site at
http://www.crcpress.com

Preface

Over the past 50 years or so, statistical analysis has evolved from being a very minor factor in scientific investigation, to being an indispensable part of almost every scientific study.

The personal computer has made it easy to apply a great variety of statistical procedures, and eliminated the computational drudgery that used to characterize this subject. So it is now very simple to use PCs in a routine fashion to get answers to statistical questions. Sounds ideal; but restricted to only eliminating the drudgery, the potential of the computer to help teach statistics is largely unfulfilled. Just cutting down on tedious computation hardly encourages users to pay attention to the assumptions underlying the statistical procedures they are using. And if the user pays little attention to assumptions, misapplications are more likely. Mindless use of computers tends to deter users from becoming familiar with how data sets reflect the characteristics of the systems they come from. This failing makes it tough to learn how to handle data properly. Finally, routine use of computers fails to help in understanding the relation between the output of statistical procedures applied to observed system data, and the basic system information being sought.

The standardized use of statistical computer programs might be okay in a production environment, but it is totally unacceptable for application in a scientific research setting. In order for scientists to choose the appropriate tools of analysis they must

- have an understanding of the assumptions under which they are operating,
- be familiar with the characteristics of data arising from different kinds of systems,
- have experience and be comfortable with what to expect when statistical procedures are applied to data with special characteristics.

This text concentrates on enabling its readers to achieve the three aims just listed. The key to our approach is the integration into our statistical analysis of a feature which is inexpensive and easily installed on just about all personal computers — this feature is the ability to rapidly generate large data sets with statistical characteristics that can be specified by the user — **a process called Monte Carlo simulation.**

With Monte Carlo techniques we can simulate just about any process we can conceive of, i.e., any one for which we have an acceptable model (description). So we can expose the reader to situations resembling those they are likely to encounter when they are trying to solve serious scientific problems. But they will have one great advantage — knowing what lies behind the data they are currently observing should give them a good idea of what to expect and how to interpret data later on.

Data sets can be generated to represent the measurements that arise from a variety of systems. They can then be examined by descriptive statistics programs, to provide a good intuitive feel for what properties these data sets can be expected to exhibit. And finally a variety of available statistical procedures can be tried out, to see how their outputs relate to the underlying characteristics of the system being examined.

With this approach we can do what otherwise can only be accomplished by relatively few with accessibility to expensive experimental equipment. And even with such access, actual systems whose properties are known are likely to be in short supply. Extensive use of Monte Carlo methods provides a stimulating way to develop the instincts leading to effective methods of data analysis, methods which get the kind of answers that are needed. This should not be surprising, because Monte Carlo simulation capability is the statistical equivalent of laboratory availability in the physical and biological sciences. And not many can get comfortable and skilled in these areas without such experience.

Of the many statistical packages that are on the market, I chose MINITAB[1] because it is simple to learn, easy to use, and widely available. One particularly powerful feature is the ease of writing **macros** based on the commands which appear in the session window; macros are short abbreviations for much of the mouse pointing and clicking needed by those with little computer experience. They allow the user to store, easily modify, and rerun long series of tedious actions with very little effort.

The organization of this text starts with a careful introduction to the aims of scientific investigation, and the nature of deterministic and statistical descriptions. This is followed by an overview of the various types of statistical problems which will be investigated in later chapters, to provide the reader with an idea of the scope of this text.

Following this, the reader is introduced to some of the accepted ways of sampling to obtain information in a statistical context. Then comes the subject of descriptive statistics, which presents some of the most useful ways of determining the characteristics of samples. As a result, the reader gains experience in learning proper ways of obtaining data, and how theoretical (population) characteristics are reflected in samples.

Chapter 4 develops the theoretical probability background which is the foundation of statistical inference.[2] Besides presenting many interesting techniques and results, this foundation is indispensable for a number of additional reasons. It provides the understanding of why statistical methods work, and criteria for judging just how well they perform. Probability theory provides the tools for determining the simulated-sample sizes needed to use Monte Carlo methods effectively, and to determine the actual sample sizes needed to extract the needed population information. It also provides theoretical results to use as a yardstick for judging the adequacy of the random number generators on which Monte Carlo methods are based.

---

1. MINITAB is a trademark of Minitab Inc. in the United States and other countries and is used herein with the owner's permission. The author is extremely grateful to Minitab Inc. for the assistance their staff provided.
2. The process of determining the characteristics of the populations from which the samples arise.

Armed with the experience built up thus far, the reader is now ready for exposure to the estimation methods, the techniques of testing hypotheses, and the regression procedures which form the latter part of this text, and are the core of statistical applications to scientific problems. The reader should then have some capabilities for

- judging the merits of and comparing estimates of the unknown quantities being sought,

- determining just how well various commonly used tests of hypotheses are performing,

- fitting curves and surfaces which describe how different variables are related to each other.

With the Monte Carlo methods to which they have been exposed, those engaged in scientific research have an additional capability which was not even envisioned prior to the availability of digital computers, namely the ability to obtain usable answers to many probability problems easily — problems that used to be beyond the reach of almost everyone.

To assist in absorbing the material being presented, a substantial number of practice exercise sets and problems have been included. The exercise sets are to ensure that facility is achieved with most fundamental methods and the models on which they are based. The problems (accompanied by many guiding hints) are used to solidify the conceptual framework of the subject. and to have the readers derive some of the useful results for themselves. Solutions to about half of the practice sets and some of the problems are provided at the end of the book. The review exercises at the ends of the chapters provide the readers an overall indication of the progress in each new area.

The choice of numbering system and some of the notational conventions used here need a few words of explanation.

Headings related to the development of the statistical concepts are numbered in one series. These headings include Theorems, Definitions, Notations, Assertions, Topics, and Problems. Exercise sets and Examples are numbered in a separately numbered series, since these relate to illustrations of ideas that have already been introduced. Equations and other expressions requiring reference are numbered separately, these numbers appearing on the right side of the pages.

Sometimes, when two items are introduced simultaneously and have much in common, what pertains to the second item is enclosed in square brackets. For instance ..*if the result is positive* [*negative*] *it corresponds to a gain* [*loss*]... . To indicate the end of an example, the black filled square is used on the right, as shown.

■

To indicate the end of the statement of a theorem or assertion, we use the unfilled square as shown.

❑

Advanced optional sections are in smaller print, and are delineated by the following pair of symbols.

⌈

⌊

Additional information will be available on the World-Wide-Web at http://www.crcpress.com. In particular, in the download section of this website, the macros given in the text will be available at no charge, to avoid requiring readers to type them in to their computers by hand. Also certain programs (such as the ones for binomial confidence limits and variance ratio confidence limits) will be similarly downloadable. Questions, criticisms, and suggestions may be sent to me by e-mail at jirosenb@utmb.edu.

There were several people whose help was indispensible — first and foremost because she solved so many of the computer problems that had me on the ropes, and came up with blunt, but very accurate criticism, is my wife, Lisa. The assistance provided by James Yanchak of CRC when no one else had a clue, simplified the production of this book substantially. Sylvia Wood, the project director, and Bob Stern the editor who suggested this book, also helped materially in eliminating many of the rough spots. Helena Redshaw was always available to help out when no one else could be found. To all of them, my sincere thanks.

Judah Rosenblatt

Table of Contents

Ch 1

Introduction

1. Scientific Method

In this text we try to explain how we go about investigating the quantitative aspects of scientific problems. Throughout history people have tried to supply explanations for what they observed. For most of this time the explanations were pretty naive — in the sense that these explanations at best supplied comfort without providing any useful control of the environment. For example, many phenomena were *explained* by inventing *Gods* — of *war, love, thunder,* the *sea,* etc. The Greeks invented the *atomic theory* without any serious attempt at developing it or obtaining any practical applications. Over the last 400 or so years we have developed the *scientific method*, which provides the most effective approach so far for construction of theories (descriptions) of the systems we want to study. Unless we can examine a system we are interested in, it's unlikely that we can learn much about it. So it seems evident that if we want to construct a useful description of such a system — one which allows us to predict and even control its behavior — we must make use of observed measurements[1] (data) both to construct the theory and to validate its usefulness. **This is the core of the scientific method**. To most of us involved in scientific investigation the merit of the scientific method is obvious. But it wasn't obvious until recently. Aristotle came to conclusions, without ever checking their validity (for example, that women had fewer teeth than men) and most of the western world took his word as gospel. Physicians in the middle ages applied a variety of treatments for various diseases without any attempt at keeping score on the proportions of recoveries. More recently Freud constructed his psychoanalytic theories unhindered by the need for reality checks, and his more dogmatic followers continue in this vein without even trying to determine the effectiveness of his methods. Even though most people at least pay lip service to the necessity for conclusions based on actual measurements, it is somewhat discouraging to realize how great a percentage of what we believe about systems ranging from automobiles to the human body is based on totally unsupported advertising hype[2].

Just how the scientific method is implemented to help solve current problems takes a substantial amount of development. This implementation involves some thorny issues, such as

1. measurements in the extended sense, which includes measurements by our senses — sight, sound (including patient reported symptoms), touch — as well as measurements made by instruments such as voltmeters etc.
2. gadgets for increasing your car's mileage that actually decrease it, useless cures, nostrums and elixirs for ailments both real as well as those invented by advertising agencies. These and many other such frauds are documented every month in Consumer Reports.

- Deciding on the initial assumptions to be made about the system
- Determining what measurements are feasible
- Settling on the goals that are desired, and on the measurements to be made to achieve them
- Learning how to make the chosen measurements properly
- Figuring out how to describe and analyze the behavior of the chosen measurements to solve the problems being investigated.

The first three issues above often involve both scientists and those concerned with analyzing the data. The fourth one is pretty much the exclusive task of the scientists — a great deal of their time and education having been spent in this area. The last task is often the principal one of the statistician, and is the concern of most of this book.

If we had an unlimited amount of time available, it might be feasible to introduce the aspect of the scientific method of concern here in a leisurely, semi-systematic, natural fashion — purposely repeating many of the mistakes that have been made by others in their investigations, and then learning how to avoid them. In fact, there have been some famous and well known teachers who have followed this approach. Properly presented it can be most effective, but even at its best it is likely to be very inefficient. For this reason although I will try to make the presentation natural, motivating each concept as clearly as possible, the halting steps and blind alleys and glaring errors that characterize science as it is really practiced will (hopefully) be omitted.

Some of my students have expressed a preference for being introduced to a list of techniques, and being told which of these techniques are best suited to some list of problems. This *recipe book* approach might be suitable if there were just a small number of well defined problems that people are likely to meet. To give a simple analogy, if we never had to do multiplications of numbers larger than 14, it might be best to have students memorize the multiplication table up through the number 14. If, however, all we knew is that we would never need to multiply numbers greater than 10,500, that corresponding solution would be hopeless.

In what follows in this text, we will try to present an approach that will let you get comfortable and familiar with the type of data you're likely to meet in real situations. This will be done by having you work with data that you yourselves will generate on the computer — data whose properties you will specify. The computer will become the experimental lab for statistics.

2 The Aims of Medicine, Science and Engineering

Most people would agree that the goal of the field of medicine is to improve human health. Achievement of this goal implies

- evaluation of people's health status — including diagnosis and identification of individuals' important risk factors
- developing treatments to cure, or at least, control, diseases — including medications, surgical procedures, physical therapy methods, etc.

- developing means of promoting healthy lifestyles — involving proper diet, appropriate exercise, etc. (preventive medicine).

There is probably less agreement on what constitutes science. Many people, including a depressingly large number of scientists, believe it is the search for universal immutable laws of nature, that is, a hunt for truth about the physical world around us. It is the pervasiveness of this belief that accounts for a great many meaningless controversies — from the quarrel between Galileo and the Catholic Church to the recent flap over the establishment of an institute at Baylor University to investigate the case for *intelligent design of the universe.* A definition of science which is nowhere near as grandiose, but which covers just about all of what scientists actually do, is that *science is* (the area of human endeavor) *concerned with developing descriptions of measurable quantities.* In particular science usually attempts to describe how such quantities are related to each other.

For instance, in physics, the simplest version of the law of gravity is a description of the relationship of the distance between a pair of objects with specified masses, and the force[1] of attraction between them. This description is imperfect if for no other reason than its neglect of electromagnetic effects. Newton's second law, $F = ma$, simply provides a description (also imperfect, since it neglects relativistic effects) relating the force, F, on an object of mass m, and the acceleration that this object exhibits.

No matter what branch of science you examine, you'll see that the problems being investigated just about always involve describing quantities which can be measured. In chemistry these quantities involve combining chemicals — e.g. how much energy is released when given amounts of sodium and chlorine combine, etc. In geology there is a great deal of interest in determining the relation between geologic measurements (such as seismograph measurements following a controlled explosion) and the amount of available oil in the region of interest. In psychology we want to determine the relationship between peoples' performance on some written test and their later performance in school or in the real world.

Sometimes, as in fields such as economics, the accuracy of the descriptions that are developed is pretty miserable. But lack of trustworthy results doesn't seem to discourage people from relying on such tools — just observe how many rely religiously on astrological predictions, economic predictions, and on polygraph results.

The last area introduced, the field of Engineering, is simple to define. It's just the use of scientific knowledge to design and construct tools and machinery to carry out specific tasks. For instance, use of scientific knowledge about the strength properties of various materials to construct buildings and bridges; use of knowledge of the behavior of fluids and of various types of motors to construct airplanes etc. That is, Engineering is applied science.

---

1. For the moment we will sidestep the tough question of how these quantities are measured

Following are some questions for those of you who are not comfortable with the views just presented. The basic issue of these problems is whether or not they can be reasonably phrased as questions involving descriptions of, or relations between quantities, which allow an agreed upon means of measurement. These questions may themselves cause some discomfort, because there may be no agreed-upon correct answers.

Problems 1.1

1. Does the question of determining your car's mileage under specified conditions allow a reasonably scientific formulation? If so how would you phrase this question in terms of agreed upon measurements?

2. Same as the previous question for whether the effectiveness of the flu vaccine allows a reasonably scientific formulation?

3. Same as question 1, concerning whether a person is dead.
    (This is a question whose answer has been changing over time.)

4. Same as question 1 regarding whether the polygraph system is an effective *lie detector.*

5. As mentioned earlier, there is a disagreement about an institute at Baylor University dedicated to investigating the issue of whether or not the universe was a product of intelligent design. Is this question a scientific one? Elaborate.

6. The controversy between Galileo and the Church is often   presented as an argument over whether *the earth revolves around the sun, or the sun revolves around the earth.* How should this controversy be resolved?
    Hints: When you say that the earth revolves around the sun,
          what type of measurements (if any) are you talking about?
          When you say that the sun revolves around the earth
          what do you really mean?
          In either case, what might you be trying to determine or
          compare?

7. In theory does it seem possible to satisfactorily resolve the question of whether the death penalty is an effective deterrent to crime? How about practically?

8. Examine some of the current controversies confronting society, to decide which of them can reasonably phrased and resolved in scientific terms

3 The Roles of Models and Data

We refer to a description of (the behavior of) a set of measurable quantities as a **model**[1] for these quantities. Models are essential in just about every phase of medicine, science and engineering.

---

1. A model is also called a **theory.**

• In medicine a model for blood pressure is needed, if for no other reason than to design instruments capable of measuring it. A very simple model for blood pressure might just specify the smallest and largest values that we expect to see.   It's reasonable to require that the meter on which blood pressures are displayed includes this range of values. A more useful model for blood pressure might also include the *normal range* for blood pressures (range of values which are desirable from a health standpoint).

• A model relating blood pressure and age might consist of the expected blood pressure at each age — or for more flexibility, might provide the *normal range* of blood pressures for each sex at each age. Such models can be major factors for use by physicians in deciding on a course of treatment.

• Aircraft manufacturers need models specifying how full their products need to be in order to break even. Lack of attention to this led to the demise of the British Vickers Viscount turboprop.[1]

• Businesses need models for demand as a function of time, in order to decide how much they should be producing as time passes. Companies that use poor models can lose lots of money (as happened with Lucent technology when they predicted that the public didn't want very high speed data transmission).

• In examining biopsies for cancer diagnosis, models are necessary to decide which of the measurements that are made correspond to healthy cells and which to cancerous ones.

The examples above should be convincing evidence that models are essential for scientific development. It follows logically that we should be interested in how to decide on a model. Two aspects of this process are worth stressing

• A model chosen to describe some specified quantities will probably have to vary with (depend on) the uses we have in mind.

• It's not likely that we'll be able to choose a model without examining some actual measurements which the model is supposed to describe.

To make a convincing argument that its uses will be important in determining an appropriate model of some quantity, let us consider mileage of a given brand of automobile (driven under specified conditions, such as the type of gas, the speed, temperature, type of road, etc.) as the quantity of interest. A model which provides the mileage accurate to 3 miles per gallon might be quite useful to the average driver in deciding when it's necessary to fill up. This same model could drive the owner of a

1. Although they didn't know it at the time, they needed to be quite a bit over 50% full to break even. This hurt airlines which had to increase capacity on a given route from one airplane (close to full) to two planes — each now less than half full, and each losing money.

fleet of taxis into bankruptcy due to its lack of accuracy. On an even more serious note, use of a model associated with the weight-loss medication Phen-Fen, which did not concentrate enough attention on effects other than weekly weight loss, led to disaster.

For the following problems, do you think it is possible to develop useful models? If not, why not? If so, what variables do you think should be examined to construct useful models for them? Explain why. (There aren't any universally agreed-on answers.)

Problems 1.2

1. You want to figure out how early to leave for work.

2. You want to know whether you can afford a Lincoln town-car.

3. A hospital wants to determine whether or not to expand its cardiac care facilities.

4. The Center for Disease Control wants to develop an influenza vaccine for use next year.

5. General Motors wants to determine whether to develop an automobile with electric brakes.

6. The government wants to decide whether to support development of a car which automates the driving process.

7. Think up some problems of your own for this set.

In order to simplify discussing how data is used in determining models, it's worth introducing the two main types of models — deterministic and statistical ones.

4 Deterministic and Statistical Models

The early models in physics, such as Newton's laws provided unique answers to questions such as "*How fast will an object dropped from 1000 feet above the earth be going after 8 seconds have elapsed?*", and "*When will the next solar eclipse occur?*". Models which provide unique numerical answers to all such numerical questions are called **deterministic**. Up until the beginning of the twentieth century deterministic models for scientific or engineering questions were usually adequate. That is, discrepancies between the models' answers and the actual measured results were usually inconsequential.   For all practical purposes of the time, prediction of eclipses was perfect. Almost all people treated with the smallpox vaccine were protected against smallpox. It has only been fairly recently that errors in the values predicted by the models we use have become important. The radar of world war II was more than adequate for the aircraft guns of the time. Today, with flight far beyond the speed of sound, even small errors in aiming can make the difference between success and failure in a battle. Our accuracy requirements have now become sufficiently stringent that discrepancies from our models must be taken into account. In addition we are dealing with increasingly complicated systems, and the models for these systems cannot feasibly take account of all of the quantities which affect the results.

For the reasons just outlined, it has become necessary to introduce models whose predictions are not only single numbers, but rather, ranges of numbers (intervals) which are more and more likely to include the value to be observed as this range becomes wide enough. So now, instead of predicting a unique mileage of, say, 20 miles per gallon, our model might yield a whole collection of predictions, examples of which might be

between 15 and 25 miles per gallon, with probability .999
between 18 and 22 miles per gallon, with probability .98
between 19.5 and 20.5 miles per gallon, with probability .42, etc.

If too narrow a range is chosen, there's not much assurance that the *true*[1] *value* will lie in this range. On the other hand, too wide a range might not be useful, even if you're virtually certain that it includes the *true value*.

Models of the type just introduced are called **statistical** or **probabilistic** or **stochastic models**.

The class of statistical models is an extension of the class of deterministic ones, since every deterministic model may be considered as a special case of a statistical model in which each range that includes the predicted answer has probability 1, and every other range has probability 0.

Problem 1.3

1. How do you decide when to use a deterministic model and when to use a statistical one.

Problems 1.4

In the remaining problems, explain why you would choose a deterministic or a statistical model.

1. The FDA wants to determine if sun-screen really protects against ultraviolet damage.

2. Medical research people want to find out the worth of cardiac bypass surgery.

3. Research people want to find out what variables are important in avoiding respiratory infections.

4. You want to decide whether to do your grocery shopping at Walmart's, Kroger's or Safeway.

5. For advertising purposes Ford Motor Company wants to determine whether its best Lincoln gets better gasoline mileage than the best Mercedes.

---

1. Here *true value* refers to the actual value that will be observed.

6. To determine an estimate of mileage of your car for purposes of trip planning, you make measurements of the mileage your car gets under the toughest driving conditions you can find.

5 Probability Theory and Computer Simulation

In section 4 we noted the dependence of statistical models on the concept of probability. Probability is necessary both to describe statistical models as well as to help decide which statistical models are most applicable. So we have to get some understanding of the probability problems we'll be encountering, and the methods that will be used to handle them. Our examples will be with coins or dice, which illustrate the basic ideas without the complex burden of reality.

The *fundamental aims of probability theory* are to compute probabilities, (or determine something about probabilities, such as approximations and properties of the error of such approximations). *Often we start out assuming we know certain probabilities, and want to compute some related probabilities.* While such problems may be conceptually easy to solve, the theoretical approach to the solution may or may not be so simple. This is illustrated in the following two examples.

Example 1.5: Toss of a pair of dice

It's common to assume that the result of tossing a pair of dice is an ordered pair (d_1, d_2), where d_1 represents the face showing up on the 1st die and d_2 that on the second die. There are 36 possible results, as seen in Table 1:

Table 1:

(1,1)	(1,2)	(1,3)	(1,4)	(1,5)	(1,6)
(2,1)	(2,2)	(2,3)	(2,4)	(2,5)	(2,6)
(3,1)	(3,2)	(3,3)	(3,4)	(3,5)	(3,6)
(4,1)	(4,2)	(4,3)	(4,4)	(4,5)	(4,6)
(5,1)	(5,2)	(5,3)	(5,4)	(5,5)	(5,6)
(6,1)	(6,2)	(6,3)	(6,4)	(6,5)	(6,6)

If each of these 36 results is assumed to have probability 1/36 (as a pair of *fair* dice should yield), then it is reasonable to assume (as we'll see more formally in CHAPTER 4) the probability that the result of tossing two fair dice will be one of any specified set, R, of possible results, is just (1/36)x(# of results in R). Hence, the probability of a sum of 11 is just 2/36 [(6,5) and 5,6) being the only possibilities that lead to a sum of 11].

This theoretical computation is rather easy.

■

Example 1.6: Toss of ten dice

The result of tossing ten dice can be represented by a sequence $(d_1, d_2,...,d_{10})$ of 10 elements, where for any i, d_i represents the result of the i-th toss. Reasoning as in the previous example, we conclude that there are 6^{10} possible such sequences of length 10 (compared to $6^2 = 36$ sequences of length 2 in the previous example). Each possible such sequence is assumed to have probability $1/6^{10}$. So the structures in this and the previous example are essentially the same; and the probability of getting a sum of 35 on ten dice tosses is thus

$$1/6^{10} x(\text{\# of sequences, } (d_1, d_2,...,d_{10}) \text{ with } d_1+\cdots+d_{10} = 35). \qquad \textbf{\textit{1.1}}$$

Because of their large number it is nowhere near as simple to determine the number of sequences in expression 1.1 as it is to determine the number of ordered pairs (d_1, d_2) with $d_1 + d_2 = 11$ (or with $d_1 + d_2$ equal to any specified value).

∎

The previous example illustrated the practical difficulty of the theoretical approach to computing a probability in a conceptually simple situation. The difficulty of such computations in more complex situations can often be an insuperable barrier. *So we'll now consider an alternative approach, based, in a sense, on what you might do if you knew no theory at all*.

Suppose you wanted to determine the probability of a sum of 11 in tossing a pair of dice. You might very well start out by actually tossing a pair of dice many times — possibly 1000 times. The proportion of tosses with a sum of 11 would seem like a reasonable estimate of the probability of a sum of 11. The main obstacles to this approach are that it is time consuming, incredibly boring, and likely to yield book-keeping errors. And, even if you had enough patience and persistence you probably wouldn't get a perfect answer. Fortunately, today's personal computers allow us to overcome the boredom and bookkeeping problems, and nonetheless obtain reasonably accurate answers most of the time. What we do is program (instruct) the computer to toss dice, (well, not *actually* toss dice, but to *imitate* the results of doing so) and let it keep track of the proportion of tosses whose sum is 11. Programming a computer to imitate an experiment with a known model is referred to as *Monte-Carlo simulation*. Most experiments — from tossing dice to playing a baseball game with known batting averages, probabilities of stealing bases, etc. — can be simulated on a computer. And while it's a bit harder to simulate many tosses of a pair of dice than it is to theoretically compute the probability of a sum of 11 on the toss of a pair of dice, the effort to simulate more complicated experiments and observe the proportions of interest is usually orders of magnitude smaller than that of theoretically computing the probabilities being estimated by these proportions. Let's see how this works using the Minitab Statistical package.

Example 1.7: Simulation of sums on dice

We presume here is that you know how to invoke this package. Here is essentially how your screen looks when you have first invoked the Minitab program.

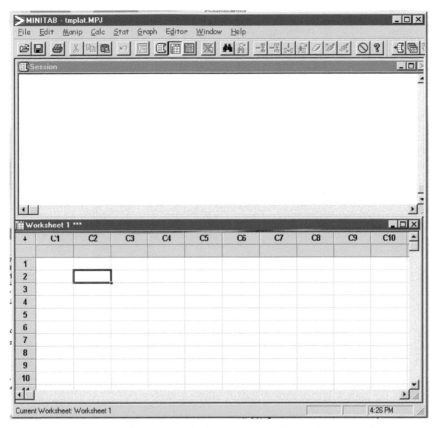

In order to have a written record of the commands that we are about to invoke using the mouse and menu bar at the top (for later use), we first left click somewhere in the Session (top) window, and then (by left clicking in the menu bar) *select* Editor (not Edit) to yield:

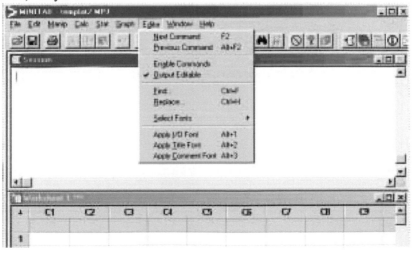

Now select Enable Commands in the above drop-down menu (by left clicking) to replace the cursor by the Minitab command prompt, MTB > followed by the cursor, |. A shown, this locates the position where typed commands appear.

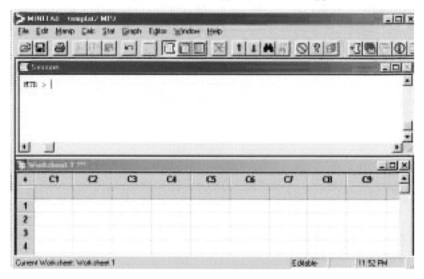

Note that the illustrated Minitab screens may change size in our text, due to space limitations. The next step is to arrange for a large number of repetitions of the toss of a pair of dice. In the menu bar click on Calc, then on Random Data in the drop-down menu, and finally on the drop-down menu which then follows, on Integer - which looks like the following (bottom a bit cut off);

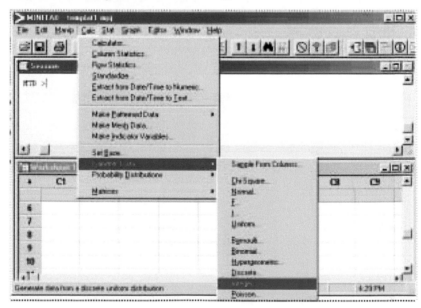

This sequence will allow you to generate data from any probability assignment which assigns equal probabilities to any finite sequence of successive integers - such as 1,2,3,4,5,6 which represents the toss of a die. Once Integer is selected, you will see the following.

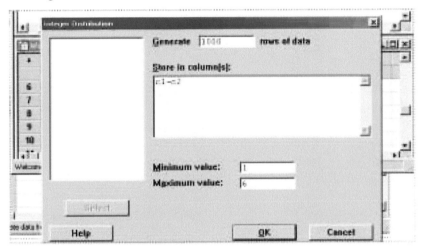

where I have filled in the values 1000 (for the number of paired tosses) c1-c2 (for the spreadsheet columns in which these results will be displayed) and the values 1 and 6 (for the possible *die values*). After clicking OK, you get the following.

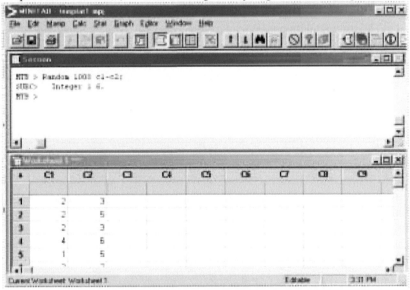

The semicolon at the end of the first line yields the *subcommand prompt*, SUBC> after you press the Enter key, while the final period (followed by *Enter*) returns you to the command prompt. Now we find the proportion of pair sums equal to 11, as shown below.

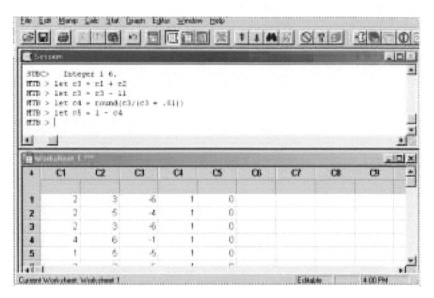

The command let c3 = c1 + c2 creates column c3 in the obvious way. The command let c3 = c3 - 11 subtracts 11 from every element of column c3.

The command let c4 = round(c3/(c3 + .01)) requires some explanation.

> Whenever a given element, q, of column c3 is not 0, the expression $q/(q + .01)$ is a number very near to 1. This expression is 0 whenever q = 0. The *round* function is defined so that *round(q)* is the integer closest to q. So columns c4 consists of numbers 0 and 1 — with 0 being put in place of 0 in column c3 and 1 being substituted for any nonzero value in column c3. This has the effect of replacing every pair sum of 11 by 0, and every other pair sum by 1.

The next command changes 0's to 1's and 1's to 0's — so that a 1 stands for a pair sum of 11. The command mean c5 (which is shown on the screen to follow) gives the proportion of 1's in column c5, which is the proportion of pair sums whose value is 11, as desired. The value, .06, is close to the theoretical value of 2/36 = .055555... which we obtained in Example 1.5 starting on page 8. For greater accuracy we would have to specify more tosses. The quantitative aspect of this assertion will be introduced when we study Binomial Confidence Intervals in Ch 5. As a rule of thumb for practical purposes when the true answer is unknown, it is common to keep increasing the sample size by some factor (such as 2) until the first such increase which no longer changes the estimate by much.

(As we'll see in the chapter to follow, this approach won't work in all cases — see problem3.7.5 on page 55.)

Note that all of this pointing and clicking required to get this result can be carried out by typing the lines that appeared on the Session Page. This will be very useful when we want to carry out modifications of a large set of operations, such as repeated simulations for tosses of more dice, etc.

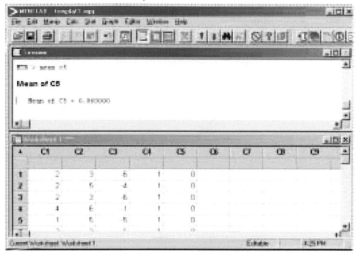

Exercises 1.8

1. Extend the computations in Example 1.5 on page 8 to tosses of 3 dice.

2. Repeat the simulation of Example 1.7 on page 9 but for 100,000 tosses of a pair of dice. What conclusions seem reasonable?

3. Check your answer to exercise 1. above via simulating tosses of 3 dice.

4. Using simulation, estimate the probability that the sum on a toss of 10 dice equals 35.

Problems 1.9.

1. How would you handle the problem of estimating all of the probabilities for the sum of a pair of loaded dice?
   Hint: On the Minitab menu bar go through the following choices.
   Calc - > Random Data -> Discrete, to get the following dialog box

Now simulate, say, 1000 tosses of a pair of dice whose probabilities $p_1, p_2, p_3, p_4, p_5, p_6$ are respectively, say, .1, .2, .3, .2, .1, .1. Prior to filling out this dialogue box, you would have to store the die values 1,2,3,4,5,6 in some column (say column c1) and their probabilities .1, .2, .3, .2, .1, .1 in some other column (sayc2). If you then filled out the box above as next shown and then click on OK, the results will be

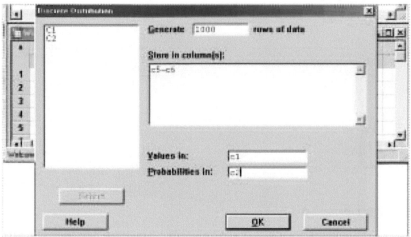

printed in successive rows of columns c5 and c6, and the command-line commands which would generate them appear in the Session window as we see below.

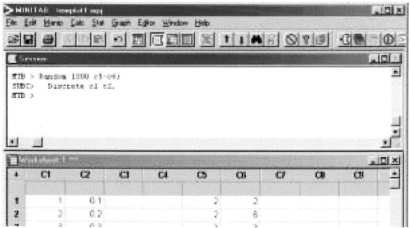

Now put the sum of columns c5 and c6 in column c7, which then has the desired loaded dice sums.   We can then proceed to estimate the probabilities of the different sum values using the same methods as in Example 1.7, starting on page 12.

2. Using simulation, estimate the probability that the sum is 7 when tossing three dice loaded as above.

3. Estimate the probability of a sum of 10 in tossing 5 dice, loaded as in the previous question.

4. Estimate the probabilities of:
       exactly 50 heads in 100 tosses of a fair coin
       at least 60 heads in 100 tosses of a fair coin
       Hints: Pretend that the Head side of a coin is labeled 1, and the
       tail side is labeled 0 — sort of like a 2-sided die.
       For the first part,
       convert the numbers in the column whose elements are the
       simulated numbers of heads in 100 coin tosses as follows.
           0-49,50,51-100 -> -1,0,1->-2,-1,0- >-1,0 ->0,1
       (that is, any results from 0 to 49 get converted to -1, 50 gets
       converted to 0, and any result exceeding 50 gets converted to 1,
       by first subtracting 50 from every result, then adding a very
       small number to every result, then dividing by the magnitude
       of the result, and finally rounding the result, giving the -1,0,1
       as was done starting in the dialogue box on page 12. Then just
       continue as in that example.
       For the last part, proceed almost as in the previous hint, but
       subtract 59.5 from each value converting the values greater than
       or equal to 60 to 1 and those less than 60 to -1, then add 1 and
       divide by 2 so that the original results greater than or equal to
       60 get converted to 1, and those less than 60 get converted to 0.
       Then take the mean of this last generated column.

Chapter 2

Classes of Models and Statistical Inference

2.1 Statistical Models — the Frequency Interpretation

In Definition 1.4 on page 7 of the previous chapter we introduced a statistical model as a description which provides answers in the form of intervals each with an associated probability. On an intuitive level, the nearer the probability of an interval to 1, the more sure we are that the result being described will be in this interval; if the probability is very near 1, we can be very sure that the result will be in this interval; if the probability is .5, we feel that it is just as likely that the answer will be in the interval as it is that the answer will lie outside this interval. This intuitive feeling is actually an adequate interpretation for most purposes.[1] But many of us (myself included) prefer an initial, somewhat more quantitative interpretation of probability embodied in the following topic.

Definition 2.1: The frequency interpretation

Let A be a collection of possible results of some experiment (such as an interval, if the experiment has numerical outcomes), and let P(A) be the probability that the actual result is one of the elements of the collection A. Suppose that the experiment is repeated independently n times — i.e., we attempt to repeat the experiment as closely as we can, with the results of any set of trials of the experiment having no influence on the remaining ones. Let $P_n(A)$ denote the proportion of the n trials for which the result is in the collection A. (This is referred to as the proportion of **occurrences of the event A**). The **frequency interpretation** is the interpretation of P(A) as the value that $P_n(A)$ gets close to as the number of trials, n, gets very large.

That is, **the frequency interpretation of P(A) is that P(A) represents the long run proportion of occurrences of the event A in independent repetitions of the experiment.**

❑

So, for a fair coin (where the probability of a *Head* is 1/2) the long run proportion of *Heads* in many tosses is .5. For a fair die, the long run proportion of 1s is 1/6, as are the long run proportions of 2s, 3s, 4s, 5s and 6s.

The *settling down* of proportions of occurrences of events (such as A) is referred to as *statistical regularity*. The phenomenon of *randomness* is essentially the same as that of statistical regularity — with randomness putting the emphasis on the variability which goes together with statistical regularity. For this reason many people think of randomness as chaos, which is not the case. The most important

---

1. In fact, with this bare bones interpretation, and the rules of probability, we can actually deduce the frequency interpretation.

aspect of randomness and statistical regularity is the *regularity* part, indicating the approach of proportions to their associated probabilities.

Statistical regularity represents an article of faith, since it can never be demonstrated to hold for any actual situation. For instance, we cannot toss a coin an arbitrarily large number of times. Any real coin would wear out after a finite number of tosses. So it is reasonable to ask why we would introduce such a concept, if it can never be verified. The answer is that the theories based on this concept often produce predictions which hold up very well in practice. We should always keep in mind that when a theory (description) is proposed and then fails to yield adequate predictions for some class of problems, it should be classified as unsatisfactory for these cases and not used there. This process of formulating and either accepting a theory as satisfactory or unsatisfactory for its intended use is a continuing one, and is an essential part of the scientific method.

To provide some *feel* for statistical regularity, we will see graphically how the **proportion of occurrences** p_n **of an event** A in n **trials** gets close to $p = P(A)$ as n gets large, using Monte Carlo simulation (Definition 1.6 on page 9). We will use a macro (an abbreviation for a sequence of Minitab and special *Minitab-macro* commands) to generate sequences $p_1, p_2, ..., p_{k2}$ for several values of p. (As we will see, it will not be difficult to avoid the large swings in value that often occur in the initial proportions, with a simple modification that will allow us to start our graph at a point somewhat down the line; this will make it easier to examine the nature of the phenomenon of statistical regularity.) Prior to running the macro we store $k2$ values from the Bernoulli distribution with parameter p in column c1. This distribution generates 1s with probability p and 0s otherwise. Then the mean (average) of any sequence of 0s and 1s is the proportion of 1s — corresponding to the proportion of occurrences of the event A. The macro, a general one to compute successive means,

$$x_1, \frac{1}{2}(x_1 + x_2), ..., \frac{1}{n}(x_1 + ... + x_n),$$

will be used for illustration of other properties later on. Its use here to illustrate the long-run behavior of proportions of occurrences of events is accomplished by the choice of Bernoulli variables (0s and 1s) placed in column c1. It is somewhat more complicated than is needed for means alone, to allow modification for other uses. This macro will be stored in the file c:\meanseq.mac. It isn't absolutely essential that all of the details of this macro be understood at this point in our development. However, everything on any line following the # sign is a comment, put in to make the code easier to follow for those compulsive readers who insist on understanding what they're doing.

⌐

   Here is the Minitab macro code for *meanseq*; this type of macro must start with the word GMACRO and end with the word ENDMACRO.

```
GMACRO
meanseq  # the template, same name as file storing this macro

do k1 = 2:k2 # beginning of loop to copy c1 into columns  c2 to ck2 —
          #k2 is a variable (storage location) which is used to store a number,
```

```
                    #to be set before this macro is run, via a command in the session
                    # window, say  let k2 = 200

let ck1 = c1 # so first  c2 = c1 (values in c1 are copied to the
                    # corresponding rows in c2)
                    #  then  c3 = c1 ... and so forth
   enddo #end of loop

let k3 = k2 + 2 # this and the next  command just sets up 2 storage
                    # locations needed for later use,  since it does not seem
                    # possible to write  c(k2 + 2) to represent column ck3 where
                    # k3 = k2 + 2.  k3 is the beginning column number where the
                    #  transpose will be put.
let k4 = k3 + k2 - 1 # k4 is the final column number of the transpose.

Transpose c1-ck2;   # command which makes  columns into rows.  This is
                    # needed because Minitab does not allow simple deletion
                    # of parts of columns, but does allow simple deletion of
                    # parts of rows (to get the successive means).  It must be
                    # terminated with ;  if it is followed by a subcommand.
Store ck3-ck4. # subcommand which puts the transposes of columns ck1-ck2
                    # into the indicated locations ck3-ck4).  It must be terminated with the
                    # symbol  .  if it is the last subcommand going with the
                    # store command,  otherwise with the symbol  ;

copy ck3-ck4 c1-ck2 # command to copy columns ck3 to ck4 into
                    # columns c1 to ck2 — not really necessary, but so that
                    # the transposed columns  are in a more visible position.
let k5 = k2 - 1
do k6 = k5:1 # beginning of loop to [erase nothing from first row],
                    #               erase last element of 2nd row,
                    #               erase last 2 elements of 3rd row,
                    #                         :
                    #          erase all but first element of last row
                    #          carried out starting with last row, so that
                    #          no non-deleted elements are moved

let k7 = k2 - k6 + 1 # as k6 goes from k5 to 1 (i.e., from k2 -1  to 1)
                    #    k7 goes from 2  up  to  k2.

let k8 = k6 +1  # k8 goes from k2  to 2 as k6 goes from k5 to 1 (i.e.,
                    # from k2 -1 to 1)

delete k8 ck7-ck2 # the delete command which erases row k8 from columns
                    # ck7 through ck2  - i.e., since  k8 goes from k2 down  to 2
                    #   while k7 goes from 2 up to k2, we see that this do-loop
                    #    leads to the sequence of commands
                    #        delete (row) k2 from columns c2 to ck2
                    #        delete (row) k2 -1 from columns c3 to ck2
                    #                     :
                    #        delete (row) 2 from columns  ck2 to ck2
# pictorially (starting from the bottom up  with x representing deletion)
#          col      c1 c2 .............................ck2
#      row 1        *  *  *  *  *  *  *  *
```

```
#       row 2       *   *   *   *   *   *   *   x
#       row 3       *   *   *   *   *   *   x   x
#        :                          :
#       row k2 - 1  *   *   x   x   x   x   x   x
#       row k2        *   x   x   x   x   x   x   x
enddo # ends loop  leaves last row with x₁ , next to last row with
```
x_1, x_2 ... 2nd row with $x_1, x_2,...,x_{k_2 - 1}$ and first row $x_1, x_2,...,x_{k_2 - 1}, x_{k_2}$

erase ck3-ck4 # gets rid of columns which are no longer needed
rmean c1-ck2 ck3 # puts the mean of each row from columns c1 to ck2 into
                    # that row in column ck3
erase c1-ck2 # gets rid of unneeded columns
let c1 = ck3 # means are copied to column c1 for ease of viewing
copy c1 c2   # to preserve original column in c2
let k10 = k2 - k9 +1 # needed because arithmetic cannot be done in a command
delete k10:k2 c1 #deletes the last k9 rows of column c1 — empty spaces
        # are filled by moving what remains in the column upward (none here)
        # Uncomment the above three commands to plot starting with the average of
        # the 1st k9 + 1 values; k9 a positive integer, which must be set in the
        # session window before running this macro, via typing, say, let k9 = 30
tsplot c1; # *time series plot* command to plot points (ordered pairs) of the form
        # (k, y_k) where y_k is the k-th element of column c1

connect. # subcommand to connect the plotted points - if this is the only
        # (and thus the final) subcommand to *plot*, it is terminated with .

ENDMACRO # Required ending for this type of macro

This macro will be run on a sequence of 100 observations from the Bernoulli distribution with probability p = .3. To do this we put these 100 observations (0s and 1s, with .3 being the probability of a 1) in column c1 via the Minitab commands:

let k_2 = 100
random k2 c1;
bernoulli .3.

Then because we stored meanseq.mac in the directory c:\, we need only type %c:\meanseq <return>. A good deal of activity will be seen in the worksheet window, resulting in the graph shown in Figure 2-1.

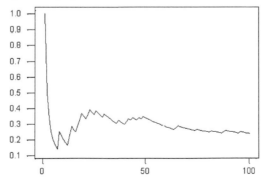

Figure 2-1 Approach of Successive Proportions to Their Probability

Let us take 400 observations to observe the nature of statistical regularity even more precisely. When we repeat the previous procedure with k2 = 400, we obtain the graph in Figure 2-2.

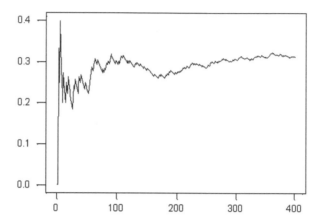

Figure 2-2 Same as Previous Illustration With 4 Times as Much Data

In Minitab,[1] this last graph took about a minute and a half to produce.

Practice Exercises 2.1

1. Run the macro *meanseq* with 400 observations and p = .5, but start the graph plot with $(x_1 + ... + x_{10})/10$ to get an expanded view of the graph of successive means. Does the approach of the sample proportions to the probability .5 seem to be *steady*? If you can, describe the sample proportions behavior more completely.

2. Run the macro *meanseq* with 200 observations and p = .5, .3, .1, .05.

   a. From the plotted graphs, what might you conjecture about the rate of approach of the sample proportions to the specified probabilities as these probabilities get close to 0?

   b. What might you conjecture about the rate of approach of the sample proportions to the specified probabilities as these probabilities get close to 1?
   Hint: *Success* occurs if and only if *Failure* does not occur.

2.2 Some Useful Statistical Models

This section appears where it does for reference purposes. Although a number of different probability descriptions will be introduced, going through them is not expected to make all of them familiar to you, much less that you will be comfortable with them all. These feelings will develop only as you use these models in a variety

---

1. Most likely my code was not as efficient as it could have been. A program to produce the same graphs, written in Splus, ran much more quickly. However, the Splus language is a good deal more difficult to learn than Minitab.

of situations. But this section should give you some idea of the breadth of situations that can be described by the field of probability.

Topic 2.2: Normal (Gaussian) distributions

We start out by looking at the most important, and most commonly encountered descriptions of them all — namely, so-called *normal* theoretical populations. Normal distributions (the statistical descriptions of normal populations) are used to describe a great variety of actually encountered populations. The adequacy of such descriptions can vary quite a bit, just as a suit off the rack fits some people well and others poorly. There are convincing theoretical reasons why certain populations are well described by normal distributions and we will meet them in Chapter 4 where the Central Limit Theorem (Theorem 4.38 on page 135) is discussed. A normal distribution is the most common description of so-called *measurement error* (the deviation between the curve fitted to data points and the data themselves) which arises when we try to fit formulas to measured data. One way to describe a normal distribution is by the familiar bell-shaped curve shown in Figure 2-3 below.[1] For any values *a* and *b*, the probability that a value drawn from the associated normal population lies in the range from *a* to *b* is the shaded area under the bell-shaped curve.

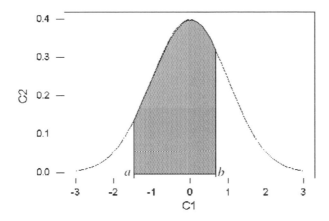

Figure 2-3 Probability of [*a*; *b*] as an Area

In general, when the probabilities for a population are given as areas as we just saw, the curve used for this assignment is called a *probability density*. The curve in the figure above is the *standard normal density*. Almost all statistical computer packages make it easy to obtain probabilities from a variety of probability densities. This has eliminated much of the drudgery that used to characterize courses in statistics.

⌈

Problems 2.3

1. Modify the macro *meanseq.mac* to handle computation of successive medians

---

1. This particular curve represents the so-called *standard normal distribution.*

(a median is a value with at least half the observations greater than or equal to it and at least half the observations less than or equal to it. It is a *middle* value.)

2. Assuming you have Minitab available, determine how to reproduce Figure 2-3 on the previous page (but without anything having to do with the shaded area). *Hints*: You first have to generate the points (ordered pairs) to be plotted. To generate the first (x) coordinates you use the menu bar as follows (first making sure you have enabled the command prompt as shown on page 10).

    Calc -> Make Patterned Data -> Simple Set of Numbers
which gets you the following dialogue box.

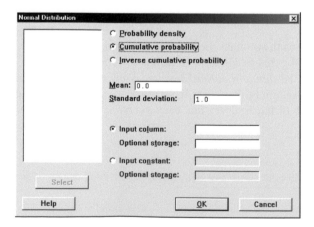

You could experiment to fill in the values above, but we will sidestep this issue and let the first value be -3, the last value +3 and the step size be .1, based on past experience. These x values will be stored in column c1. After filling in these values, click OK and you will see the first coordinate values filling column c1.

To generate the corresponding 2nd coordinates (y values) again we go to the menu bar with:

    Calc -> Probability Distributions -> Normal
to obtain the dialog box which you see below. In this box we select Probability density. The standard normal density has mean 0 and standard

deviation 1 (later we will see what these terms represent). The input column is where the first coordinates of points are — and this was c1. The *optional storage* is where the second coordinates of plotted points go, and we will use c2. Now click OK, and observe the values being placed in column c2. To plot these points we use the menu bar via:

      Graph -> Plot.

We have filled out the dialog box that resulted, as shown. For a plot with the

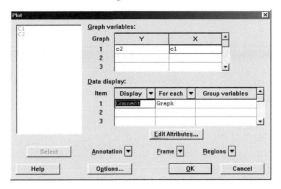

points connected, we used the drop-down menu to the right of Display, and chose Connect. After clicking OK, the desired graph is displayed. It can be printed, or stored in a file in a variety of formats.

L

Example 2.2: Computing normal interval probabilities

To actually compute the probability of a standard normal variable being in the interval from *a* to *b*, we first put the values *a* and *b* in some columns, say c1 (just by moving the mouse and clicking in the first cell in column c1, then typing the value of *a*, and then doing the same for the value of *b* just below). The values 1.1 and 2.3 were put in c1. Then from the menu bar we invoke:

      Calc -> Probability Distributions

and choose Normal, yielding the following dialogue box; we filled it as shown.

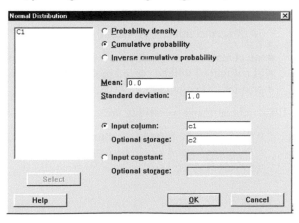

After clicking OK the Session and Worksheet windows will look as follows.

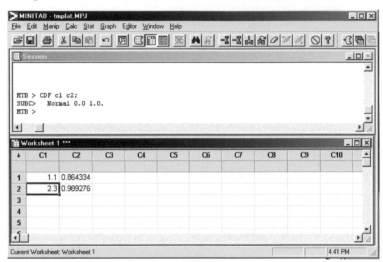

This results in putting the standard normal probability of a value less than 1.1 (the *standard normal cumulative distribution function* evaluated at 1.1) in the first row of column c2, and the probability of a value less than 2.3 in the second row of c2. Since these probabilities are respectively the area under the standard normal density to the left of 1.1, and to the left of 2.3, the difference — (second row of c2) – (first row of c2) — is just the area under the standard normal density between 1.1 and 2.3 — i.e., the standard normal probability of observing a value between 1.1 and 2.3.

After you have typed in the values 1.1 and 2.3 into column c1, all of the remainder of the computations can be abbreviated as a *macro* (a sequence of Minitab and Minitab-Macro commands, produced as a text file by any word processor or editor) as follows.

GMACRO   # Macros of this simplest type must start
               # with the word GMACRO

normintervalprob # name to indicate calculation
                  # of normal interval probability

CDF c1 c2;   # the cumulative distribution values at the elements
                # of column c1 will be placed in the corresponding
                # positions of column c2

normal 0.0 1.0. # subcommand specifying the standard normal distribution

copy c2 k1 k2   # this copies the rows of c2 into the constant storage
                  # locations k1 and k2. There must be as many constants
                  # as values in c1

let k3 = k2 - k1

print k3

ENDMACRO   # Macros of this type must end with the word ENDMACRO

If this macro is stored in c:\normintp.mac, then to execute it, just type

%c:\normintp <Enter>

Problems 2.4

1. Compute the probability of getting an outcome between 1.1 and 2.3 for the following normal distributions

   a. mean 0 and standard deviations 4,8 and 12

   b. mean 1.1 and standard deviations 4, 8 and 12

   c. Try to determine the meaning of the mean and standard deviation of a normal distribution from the results of Problem 1. above. To help draw your conclusions, graph the densities which are specified in the previous problem (see also Problem 2.3.2. on page 25).

Example 2.3: Density histograms and probability densities

Now let's see how a *density histogram* based on a random sample of size 1000 from the standard normal distribution compares with the standard normal density. A **density histogram** of a sample **is a bar graph, where the area of any bar is the proportion of observations in the sample lying in the base interval of that bar**. Of course we expect the density histogram to resemble the probability density from which the sample was drawn — but the question is *how closely?*

To get some idea we'll first draw the random sample of 1000 *standard normal* observations. (Do this using the Random Data command as we did on page 10, but select 1000 observations and choose the normal distribution with mean 0 and standard deviation 1. This could also be done with the line command random 1000 c1.) We would like to overlay the density histogram of the data in column c1 and the standard normal density. This is a trifle more complicated in Minitab than you might expect; because you need to keep the x axis scales the same for the two graphs. To accomplish this, we need to find the smallest and largest values in the random data set (either by sorting column c1 or using the line commands min c1   and   max c1). Then we set up column c2 to be the first coordinate values for the density plot (as done in Problem 2.3.2. on page 25) using min c1 as the first value, and max c1 as the last value, and, say, steps of .01. Then generate the second coordinate values of the standard normal density, also as done in Problem 2.3.2. The final step is plotting the two graphs overlayed. Here, in abbreviated form is how this is done. From the menu bar above the Session screen choose

Graph -> Histogram and fill out dialog box as shown on the page following.

Then choose Annotation -> Line and fill out as shown, click OK, and choose

Frame -> Multiple Graphs   and Overlay Graphs and Distinct X and Distinct Y click OK, then

Options -> Density and Midpoint and Number of Intervals 75   and click OK.

Here is the box resulting from Graph -> Histogram.

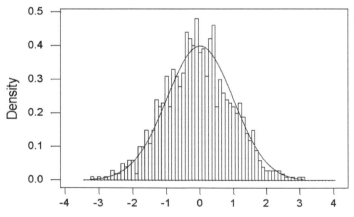

Next, click on Annotation, choose Line from the drop down menu, fill out the ensuing dialog box as shown below (using the drop-down menus to the right of the

top labels to see how to proceed) and click OK, getting rid of this last dialog box. Then click on Options in the first dialog box, follow the previous directions click OK. Now click on Frame in the first box and continue with the previous directions, until repeatedly clicking OK eliminaates all dialog boxes. are gone. You will then see the overlay in Figure 2-4.

Figure 2-4   Density and Typical Histogram

This shows how close the density histogram is to the theoretical density when the sample size is 1000. The discrepancies between these two graphs are due to two factors. First, by their very nature the histogram values, which are constant on each of the intervals chosen for the histogram, cannot be expected to equal the varying functional values of the probability density (at best, each bar height would equal the area under the density divided by the base length of the bar — that is, the bar height would be the average area over the base interval of the bar). More important is the statistical nature of the sample causing the much larger discrepancies you observe in Figure 2-4 — referred to as *sampling error*. Other random samples would yield different sampling errors.

■

Topic 2.5: Binomial distributions

A binomial distribution models the number of successes in n independent repetitions of an experiment whose probability of success is some value p in the interval from 0 to 1. The formulas for the probability of *exactly k successes* and *less than or equal to k successes* in n such trials are needed for theoretical investigations. At the basic applied level of the material being introduced here, writing down these formulas would serve no useful purpose, since these formulas convey no information that is likely to be of use at this level. To get some feeling for how the binomial probabilities vary, it is worthwhile to graph a few of the binomial distributions. We graph the binomial probabilities for four cases: $n = 7, 25$ $p = .5, .8$, omitting details, because of their similarity to those leading to Figure 2-3 on page 24.

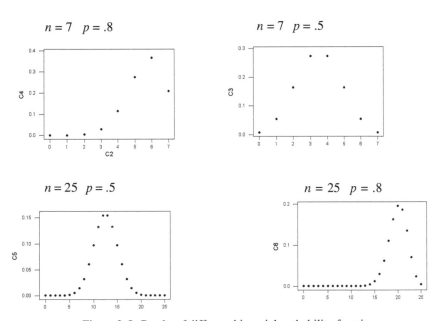

Figure 2-5 Graphs of different binomial probability functions

The resemblance of the binomial probability graphs to normal probability densities when the sample size is large turns out to be no accident (see Section 4.8 on page 133, The Central Limit Theorem and its applications).

Practice Exercises 2.4

1. First guess, and then compute the probability of exactly 50 successes in 100 experimental trials, each of which has probability .5 of success, via
   Calc -> Probability Distributions -> Binomial.

2. First guess and then compute the probability of more than 60 successes in 100 experimental trails, each of which has probability .5 of success, via
   Calc -> Probability Distributions -> Binomial
   (this time using the Cumulative probability option).

3. Calculate individual and cumulative binomial probabilities for some other values of p and other sample sizes of your choosing.

Topic 2.6: Poisson distributions

Poisson distributions are used to describe such quantities as

- the number of telephone calls initiated in a given time period
- the number of jobs arriving at the central processing unit of a computer in a given time period
- the number of automobile collisions in a day, week or year
- the number of cells with DNA damage after exposure to ultra-violet rays for some given time period.

Their first known use was to statistically describe the number of Prussian soldiers kicked by a horse over some period of time. They are also used to approximate cumulative binomial probabilities[1] when

- the sample size, n, is large,
- the product np^3 is much smaller than 1.

Practice Exercises 2.5

Graph the Poisson probability function for various values of its mean.

Topic 2.7: Uniform distributions

Uniform distributions have a probability density function whose graphs look as shown in Figure 2-6, which illustrates the uniform density on the interval $[a\,;\,b]$.

Figure 2-6 Graph of uniform density

1. See Uspensky "Introduction to Mathematical Probability", McGraw-Hill, 1937, page 135 for details regarding the Poisson approximation to the binomial.

The uniform density on the interval [-.0005; +.0005] is often used to describe the error made in *rounding off* a result to 3 decimal places; i.e., replacing the result by the closest value possessing 3 digits to the right of the decimal point. As such, it is useful in determining the error that accumulates in computations in which only 3 decimal places are carried. The uniform distribution on the interval $[-1/2^n; 1/2^n]$ (where n - 1 is the number of figures carried to the right of the decimal point in scientific [exponential] notation) is useful in determining the behavior of the error accumulated in computer computations.

Topic 2.8: Exponential distributions

Exponential probability densities are used to describe the time lengths between

- initiation of successive telephone calls,
- successive bus arrivals,
- successive electronic failures (say, due to lightning strikes),
- successive stages in certain biological processes.

Exponential distributions are completely determined by their mean (the value that the average of a large number of repeated observations is close to). In the first situation listed just above, this is the mean time between initiation of successive phone calls. Because exponential distributions are determined completely by a single quantity, they are referred to as a **one-parameter family**.

Exponential distributions have a so-called *memoryless property*, meaning that the chance of continuing for another t units of time, given that the process has been going on for T units of time already, is the same for all T. This may seem very strange — almost unbelievable, since if the exponential distribution is used to describe the time to failure of some electronic device, it would appear that the device is not wearing out. But this might very well be the case if failure was due to external causes, such as lightning strikes.Chapter 4

Problem 2.9

Graph some exponential densities and their corresponding histograms.

Topic 2.10: Weibull distributions

The class of Weibull distributions is a generalization of the set of exponential distributions. It includes distributions which do not have the *memoryless property* displayed by exponential distributions.

Topic 2.11: Gamma distributions

The process of cell division is sometimes modeled as a succession of identical *exponential* processes, all of whose durations do not have any effects on each others' durations. If we denote these durations by $T_1, T_2,..., T_n$, then the time duration taken for cell division is $T = T_1 + T_2 + ... + T_n$. The variable T has a Gamma distribution. The collection of Gamma distributions is a two-parameter family, the second parameter arises from n, the number of stages needed for cell division.

Topic 2.12: Negative binomial distributions

The negative binomial distribution with probability of success, p, and number of successes n is concerned with independent trials, each having probability p of success, which continue until exactly n successes have been achieved. The negative binomial distribution is the distribution of the number of failures occuring prior to the achievement of the n-th success. Currently this does not seem to be available in Minitab. Some use the total number of trials (rather than the number of failures).

Problem 2.13

Write a macro to produce a specific (n,p) negative binomial distribution.

Topic 2.14: Hypergeometric distributions

Hypergeometric distributions describe the behavior of a sample from a finite population consisting of 2 types of objects (say men and women, or red suits and black suits [from a deck of playing cards], or good units and bad units from some manufactured lot).

With the distributions just listed it is possible to develop many other ones. For instance, the heights of men might be described by one normal distribution, and those of women by another. The distribution of the population consisting of both the men and the women would be the so-called *mixture* of their separate distributions, and might have a density looking as shown in Figure 2-7 below.

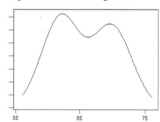

55   65   75

Figure 2-7 Graph of Mixture of Distributions

The *student t, Chi-Square, F, Logistic,* and *Beta* arise largely in determining the behavior of statistical tests and estimates, and will be introduced when these topics are taken up.

Practice Exercises 2.6

1. Solely from the density histogram of a sample of size 1000 with the uniform distribution on -.5 to .5, guess how to describe the uniform density on this interval.

2. Determine the effect on the figure that results from the effect of greater sample sizes (twice as many, ten times as many, etc.) and more intervals (100, 150,...) in the computations leading to Figure 2-4 on page 29.

3. Become familiar with the Poisson distribution probability functions and corresponding histograms for various sample sizes and population means. The Poisson family of distributions is widely used to describe the number of occurrences of such events as being struck by lightning and failure of a given type of electronic component.

4. Examine the Cauchy density, and related histograms for sample sizes of 10 with 5 intervals, 20 with 8 intervals, 200 with 15 intervals and 2000 with 75 intervals.   Do your results look strange?   The Cauchy distribution arises from a spinning pointer   having a probability of falling in any arc of radian angle ϑ being proportional to this angle. The Cauchy variable, X, is the value pointed to by this pointer, as shown in Figure 2-8 .

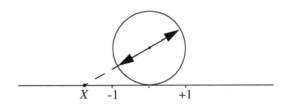

Figure 2-8 Model for the Cauchy Distribution

The Cauchy probability density can be shown to be a *bell-shaped* curve For those of you without access to Minitab, here is a density histogram generated by a Cauchy sample of size 200. What can you do to make it look more like the density it should resemble?

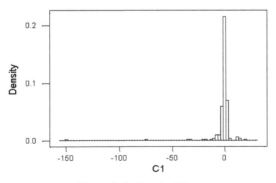

Figure 2-9 Cauchy Histogram

5. To get a feeling for their behavior, become familiar with the densities, probability functions and related histograms of other distributions available in Minitab. If you feel ambitious, overlay histograms and

related densities. In particular, determine how large a sample size and how many intervals are needed so that the histogram closely approximates the probability density. Use your experience to determine a reasonable rule of thumb enabling you to conclude that you have enough data that, with the proper number of intervals, the histogram gives you a good picture of the probability density. (To overcome the Minitab maximum of 100 intervals, do not use the *number of intervals* option. Instead, take the *midpoint/cutpoint positions* choice, e.g., -3 : 3 / .04 will give you about 150 intervals.)

2.3 Narrowing Down the Class of Potential Models

We saw on page 8 that the fundamental aim of probability theory is to compute something about a probability — often from knowledge about some other related probabilities — as illustrated for a pair of dice in Example 1.1 on page 8. Some other, more substantive examples of probability problems are:

- Determining the probability distribution of the total amount of time for some manufacturing process from the probability distributions of the times for its separate stages.

- Determining the probability distribution of the largest of a sequence of measurements from their individual probability distributions. (Such results can be important in designing reliable equipment.)

- Determining the probability that NASA space ship launches will have a perfect record over a ten year period, knowing the probability of success for each individual launch.

- Calculating the probability of a run of at least k successive failures in repeated trials whose probability of failure is q; this problem is of interest in quality control where such a run leads to readjusting the manufacturing process — the answer tells how likely it is that when the process is actually under control, it will needlessly be readjusted. It is also of interest in determining the probability of losing in the *double your money when you lose* strategy of betting.

- When trying to determine the validity of some scientific hypothesis by means of some data dependent procedure, it is of importance to determine the probability of rejecting the hypothesis of validity of the hypothesis under various conditions such as, that the hypothesis is valid, or that the hypothesis departs from validity by some given amount. That is, testing statistical hypotheses leads to many distinct probability problems.

- Determining the distribution of particles emitted from manufacturing facilities under specified weather conditions, from probability assumptions about the nature of the dispersion process in short periods of time. Computations like this are needed for pollution control.

- Determining the distribution of payout resulting from airline overbooking, knowing the number of passengers booked and the probability of no-shows for passengers.

- Determining the distribution of auto insurance casualty claim payouts from an assumed distribution of accidents and injuries.

In each of the problems just listed, a single model had been chosen, allowing the computation of the desired probabilities from the ones assumed known.[1] It is reasonable to ask how certain probabilities can legitimately be known; and there are several possible ways that this can occur. In a few cases the physical situation might be such that the assumed probabilities seem *self-evident* — for instance in the toss of a symmetric coin it is reasonable to assume a 50-50 chance of heads; similarly for a pretty symmetric die the assumption that all faces are equally likely seems tenable. The assumption of uniform roundoff errors seems plausible in many computer problems. But even in these cases, it is wise to try to verify the suitability of such assumptions before betting the farm on them.

For the most part, at the start of the great majority of scientific investigations, we rarely start out with but one model. At best we usually have a collection of *candidates*, containing (we hope) at least one model good enough for the purposes we have in mind. Our interest is usually in observing some relevant data which can be used to *narrow down* this collection of potential models — sometimes narrowing down to a collection which for all practical purposes behaves like a single model, and at other times to just a more manageable collection. Some examples should clarify the desirability of these goals.

Example 2.7: Election poll

To make this illustration simpler, let us suppose that we know who will vote, that it is a two-person race, and that the popular vote winner is the one who gets elected. Let p be the proportion of all voters who will vote Republican in the actual election. By appropriately[2] choosing and examining a sample of voters, we would like to narrow down to one of two cases — namely $p < 1/2$ (Democrat wins) or $p > 1/2$ (Republican wins).[3] We will assume that a natural choice for the original collection of potential models for the number of Republican voters in the sample is the set of binomial distributions with sample size n (= total number of voters in the sample) and probability p with $0 \leq p \leq 1$. By examining the sample, we want to narrow down to one of the two classes:

---

1. As will be seen in Chapter 4, although in theory the desired probabilities can be determined, the actual theoretical computations can be quite daunting.
2. How to accomplish this appropriate choice will be discussed in Chapter 3.
3. For simplicity we will arbitrarily exclude the possibility that $p = 1/2$. After all, we're not asking for a perfect description.

the binomial distributions with sample size n and probability $p < 1/2$
the binomial distributions with sample size n and probability $p > 1/2$.

We will see in Section 6.3 of Chapter 6, how large the sample needs to be to obtain an acceptable solution to this problem (see Exercise 6.12 on page 208; from the *Stat* drop-down menu choose *Power and Sample Size*, then 1 *proportion*,and finally fill in *Alternative values of p* and *power values.*)

■

Example 2.8: Estimation of physical or biological quantities

Many a scientific problem is motivated by the desire to determine the value of some quantity μ, of interest (such as the speed of light, the strength of some new material, or the level of expression of some gene [say the rate at which the gene produces a needed protein]). If the observed variability in the measurements of μ arises mainly from a well-behaved measuring instrument, then one reasonable class of models might postulate repeated independent measurements $X_1, X_2, ..., X_n$, each normally distributed with unknown mean μ and known standard deviation σ (recall that a small standard deviation indicates low variability). Initially we may feel that we know nothing at all about μ, so that for the problem being considered, μ could be any positive number. To narrow down the class of models from the large class consisting of all normal distributions with positive mean and standard deviation σ, we make use of the theory of confidence intervals (which is developed in Chapter 5). A **confidence interval** for some unknown quantity is an interval constructed from observed data, and which includes this quantity with at least some specified probability (often .95 or .99).

It will be shown in Example 5.5 on page 155, that under the above assumptions, a .95 level confidence interval for μ based on $X_1, X_2, ..., X_n$ ranges from

$$\frac{1}{n}(X_1 + ... + X_n) - 1.96\frac{\sigma}{\sqrt{n}} \quad \text{to} \quad \frac{1}{n}(X_1 + ... + X_n) + 1.96\frac{\sigma}{\sqrt{n}} \qquad 2.1$$

So with the aid of the measurements $X_1, X_2, ..., X_n$ we can reduce the original class of candidates for describing the measurements (normal distributions with unknown positive mean, μ, and known standard deviation σ) to the class of normal distributions with mean μ in the range specified by Expressions 2.1 above. Of course there's a price to pay — namely that we are only 95% as confident that this smaller class of distributions contains what we are looking for, than we were about the original larger class. Nonetheless, the trade-off is worthwhile.

■

Topic 2.15: Distinguishing characteristics of statistics

In statistical problems we do not have just a single model to describe our measurements, but rather a class of models which is assumed to include at least one suitable model.

The object of statistics is to make use of data to *cut down* this class of models in some legitimate manner, to one more suited to the goals we have in mind.

The remainder of this section consists of examples illustrating the scope of statistical investigations. We start with the simplest approach to finding how variables are related.

Example 2.9: Simple linear regression

The task of determining how two important variables are related has been of interest ever since scientific method was developed. One of the earlier examples of such relationships is Newton's second law of motion, $F = ma$ (relating acceleration, a, to force, F) while one of the more recent ones is Einstein's $E = mc^2$, relating the energy, E, available from an object, to the amount of mass, m, of that object.

The most commonly used type of deterministic model relating two variables y and x is of the form of a line

$$y = ax + b,$$

where a and b are unknown numerical constants. The main reason that so many two-variable investigations begin with this class of models[1] is that, at least over a short range of x values, the plotted graphs of experimentally observed (x,y) pairs frequently look close to lying on a line. But due to experimental error,[2] there is usually no line on which a substantial number of such points lie, even if the underlying mechanism generating these points was a straight line. A common statistical class of models, accounting for this phenomenon is given by

$$y = ax + b + \varepsilon, \hspace{3cm} 2.2$$

where ε (representing the effect of experimental error) is a normally distributed random variable with 0 mean and unknown standard deviation σ, and as above, a and b are unknown numerical constants.

At the start of an investigation using this type of model, it is common to ask whether y really seems to depend on x — by using data to *test the hypothesis that $a = 0$*. The approaches to answer this question will be investigated in Chapters 5 and 6 . Note that either conclusion *that $a = 0$*   or   *that $a \neq 0$* would narrow down the class of models being considered.

Actually this formulation is somewhat oversimplified, because sometimes the answer must hedge by concluding that the evidence to exclude the possibility that $a = 0$ is inconclusive, rather than concluding that $a = 0$. This unfortunate situation can occur when we have not observed enough data. But concluding that $a \neq 0$ **would** narrow down the class of models. Also it may be the case that $a \neq 0$, but that a is so close to 0, that for practical purposes it should be considered 0. Probably the simplest way to deal with these issues is by means of confidence intervals, to be discussed in Chapter 5.

---

1. The class of all models of the form $y = ax + b$ for real a,b.
2. Which can be thought of as variability induced by factors not under the experimenter's control, or variability introduced by the measuring instrument.

We might also be interested in obtaining confidence intervals for any of the unknown constants, $a, b,$ and σ — which we know would also constitute a reduction of this class of models.

■

Example 2.10: Multiple linear regression

A natural extension of the previous example occurs in trying to use data to find something about the relation between k quantities which are immediately available (so-called *independent variables*) and a single quantity whose value may not be available (the dependent variable). The type of model most commonly used is of the form

$$y = a_1x_1 + a_2x_2 + ... + a_kx_k + \varepsilon.$$ 2.3

Here for $i = 1,...,k$, x_i is the (assumed known) size of independent variable i; ε (representing experimental error) is assumed to normal with mean 0 and (likely unknown) standard deviation σ. The quantities a_i, called *regression coefficients,* are the responses per unit amount of variable i.

The sum $a_1x_1 + a_2x_2 + ... + a_kx_k$ is called the **deterministic part** of the model, and the quantity ε its **random part**. The quantity y is called the **dependent variable**.

The independent variables might be *temperature, pressure, humidity* in some metal manufacturing process and the dependent variable might be the strength of the final product. Statistically, we might want to determine which of the regression coefficients (if any) appear to be 0, or we might want confidence intervals for some of the regression coefficients, to get an understanding of how each independent variable influences the dependent variable.

This class of models can represent a much greater variety of relationships than you might imagine. For instance, the deterministic part of the model can represent any k-1st degree polynomial in the independent variable time — simply by letting x_1 be 1, x_2 be t, x_3 be $t^2,...,x_k$ be t^{k-1}. In such a case, to test whether the quantity y does or does not change with time, you could test the hypothesis that all of the regression coefficients except for a_1 are 0.

Confidence intervals for any set of regression coefficients and for the standard deviation σ could also be constructed. As an extension, *confidence bands*[1] for the entire deterministic part over any finite time interval can be constructed.

■

Example 2.11: Comparing two treatments —*t*-tests

Two samples are chosen from some population to be as similar as possible with regard to how individuals in these samples react to two treatments for some condition they have (e.g., to arthritis). The object is to determine which of two treatments is

---

1. A *confidence band* is a strip which has a specified confidence of including the graph of the given deterministic part.

superior for treating their condition. Hopefully, each treatment would affect both samples in the same way. The first treatment is applied to those in the first sample and the second treatment to those in the other sample. Let us suppose we have some numerical criterion for how the patient reacts to either treatment, with *smaller* (less pain) being better. The form of the model is chosen to be

$$y = \mu_i + \varepsilon \quad i = 1, 2 . \qquad\qquad 2.4$$

The notation here needs some explanation. Here μ_1 and μ_2 are two unknown constants. For $i = 1$ or 2, the reaction of patients to treatment i is $\mu_i + \varepsilon$, where ε is a normal random variable with mean 0 and possibly unknown standard deviation σ.

In the jargon of hypothesis testing the most common process of *narrowing down* the class of models is

testing the *null* hypothesis that $\mu_1 = \mu_2$

against the alternative that $\mu_1 \neq \mu_2$

(meaning that we will be cutting down the class of models to ones of the form given by Equations 2.4 in which either $\mu_1 = \mu_2$ or $\mu_1 \neq \mu_2$).

In this situation it is becoming increasingly more common to accomplish the aims being sought by obtaining a confidence interval for the quantity $\mu_1 - \mu_2$, since this provides more usable information than the testing hypotheses format.

In cases where we know that treatment 2 is at least as good as treatment 1, we may instead

test the *null* hypothesis that $\mu_1 = \mu_2$

against the alternative that $\mu_1 > \mu_2$

(meaning that we are assuming that $\mu_1 \geq \mu_2$ and are reducing this class of models to a class of the form in Equations 2.4 in which either $\mu_1 = \mu_2$ or $\mu_1 > \mu_2$) etc.   ∎

Example 2.12: One-way analysis of variance (ANOVA)

This is an extension of the *t*-test of the previous example to more than two treatments. The model here is written

$$y = \mu_i + \varepsilon \quad i = 1, 2,...,k \qquad\qquad 2.5$$

where the μ_i are unknown non-negative constants.

The default narrowing down of most statistical packages is into the class of models of the form given by 2.5 above

for which $\mu_1 = \mu_2 = ... = \mu_k$

and the class

for which it is false that $\mu_1 = \mu_2 = ... = \mu_k$.

Confidence intervals can also be constructed for the μ_i and for the differences $\mu_i - \mu_j$ $(i \neq j)$.

■

Example 2.13: Purely additive two-way ANOVA

Suppose several different treatments for the same condition (say high blood pressure) are given to each patient in a study, with plenty of time taken between different treatments given to the same patient. We would expect that the responses of patient 1 to treatments 1 and 2 would be closer than the response of patient 1 to treatment 1 and patient 2 to treatment 2, because of the significant factors which distinguish two people. We let μ_i $i = 1, 2, ..., I$ be the average increase in effects of the I treatments above that of treatment 1(the treatment arbitrarily chosen as the *control treatment*) and s_j $j = 1, ..., J$ be the increase in the patient (subject) average affect above that of the subject 1 (the arbitrarily chosen *control subject*).[1] Let m be the average overall effect of the control treatment and subject effects. Then the model often used is

$$y = m + \mu_i + s_j + \varepsilon.$$ 2.6

In most cases it is only the μ_i which are of interest. So why include the subject effects? One reason is that this model seems more valid than the one of Example 2.12 for a problem in which multiple measurements are made on each subject (a so-called *repeated measures* model), since the one-way ANOVA model takes no account of the likelihood that measurements within a patient are likely to vary less than measurements from patient to patient). Looked at another way, this model removes the patient to patient variability from the error term ε, thus allowing more accurate estimation of the treatment effects that we care about.

The shrinking of the class of models can include tests of hypotheses about the treatment effects, or confidence intervals — for the treatment effects, or for the differences between treatment effects.

■

Example 2.14: Estimation with prescribed precision

There are many cases in which confidence intervals will be of no use if they are too long. For example a .95 level confidence interval equal to [.8; .9999] for the probability of a successful space launch would leave us with little incentive to implement the launch.

It turns out that the key to obtaining confidence intervals which have a pre-assigned length lies in knowing an *upper bound*[2] for the standard deviation of an

---

1. This is required because it is not mathematically possible to estimate absolute treatment and absolute subject effects, but only their relative effects — since any uniform rise in treatment effects could be offset by an equal size fall in subject effects, in the sense that the distribution of the observed data would not be changed.
2. An upper bound for the standard deviation is a quantity which is greater than or equal to the standard deviation.

estimate whose theoretical mean is the quantity of interest; because then you know a value which limits the estimate's amount of variability. Then, as we will see, we can determine how many repeated observations on this estimate are needed so that the average of sufficiently many repetitions has small enough variability. For this reason confidence intervals and tests of hypotheses for standard deviations of normally distributed observations are of some importance.

■

Example 2.15: Checking ANOVA assumptions

One of the assumptions made for one-way analysis of variance (see Example 2.12 on page 40) is that all observations have the same standard deviation. If this assumption is seriously violated, the results claimed for this procedure may be way off. Hence, it is worthwhile to be able to use data to test hypotheses concerning the ratio of standard deviations associated with two different sets of observations.

Minitab has tests of hypotheses available for checking on the equality of standard deviations. Confidence intervals for the ratio of standard deviations arising from different sets of normally distributed observations will be available from a program on the crcpress.com web site. If the confidence interval for such a ratio fails to include the value 1, then you may conclude that the two standard deviations are not equal. If this interval is sufficiently short, and includes the value 1, then for practical purposes the standard deviations may be considered equal.

■

Problems 2.16

1. Referring to Example 2.7 on page 36 on election polling, suppose that a sample of 1001 people is polled, and that the Republican is declared the winner if a majority of the sample opt for him/her.

    a. For $p = .46, .47, .48, .5, .52, .53, .54$, compute the probability that based on this sample, the Republican is declared a winner.

    b. You should recognize that the procedure just described constitutes a test of hypotheses, and that the computations of part a. provide a good indication of how well this procedure works. How well does this procedure seem to work?

    c. What sample size is needed to be 95% sure of making the correct decision as to the winner, both when $p = .49$ and when $p = .51$?

2. Suppose that we have a measuring instrument for measuring *a.c. line voltage* whose error is normally distributed with mean 0 and standard deviation of 4 volts. Assume line voltage to be constant.

    a. How many observations are needed to estimate the line voltage to within 2 volts with confidence .95 via a .95 level confidence interval for the line voltage. Explain why a confidence interval is an appropriate means of obtaining the desired estimate.
    Hint: The confidence interval should be 4 volts long; see Example 2.8 on page 37.

b. If a line voltage which differs from the *nominal* voltage (115) by more than 10 volts could cause malfunctions, what is a *reasonable* length .95 level confidence interval for the line voltage to get you the needed information about malfunctions?

3. Suppose that measurements on automobile mileage are obtained by driving a car for 20 miles at 35 miles per hour, and using as the mileage estimate the quantity 20/g where g is the measured gasoline used for this test. Assume that such measurements are normally distributed with a theoretical mean equal to the true mileage and unknown standard deviation σ. Suppose that for each automobile being tested 6 trials are carried out.

a. Generate data representing these trials for two auto models, one with a true mileage of 23 miles per gallon, and
$$\sigma = 2.9 \text{ miles per gallon}$$
the other with a true mileage of 19 miles per gallon and
$$\sigma = 4.2 \text{ miles per gallon}$$
placing this data in columns c1 and c2 if you are using Minitab. Using the two-sample t-test (see Example 2.11 on page 39) determine whether the chosen sample sizes appear to be adequate to detect the true mileage differences? (Use the default choices, and the supplied confidence intervals to answer the question. For now, don't worry about the rest of the output.)

b. If you wanted to detect such a difference, how large would you take the sample sizes? (Justify your answer)

4. Since a model is just a description, provide an example of how a one-way ANOVA and a simple linear regression can be used to try to answer the same question.

5. Below is a graph of data (with experimental error) generated for a simple linear regression situation.
The points on the *true line* were generated from the
  Calc -> Patterned Data -> Simple Set of Numbers commands
  (putting data in c1) together with let $c2 = 3*c1 -2$ (so that c1,c2 are
  respectively the first and second coordinates on the true line)
The *data* were generated from the
  Calc -> Patterned Data -> Arbitrary Set of Numbers commands
  (8 numbers were generated) put in c4 (for the first coordinates).
The second coordinates of these points (but without the error)
  were generated by
        let $c6 = 3*c4 -2$ (same formula as for the true line)
The error was generated by random 8 c7;
                                normal 0 .7.
The second coordinates of the data (with error) were generatedby
        let $c9 = c6 + c7$
So, the true line coordinates are in c1 c2 and the data with error were in c4 c9.

The plot shown below was most easily done by the following session commands:

    plot c9*c4 c2*c1;
    symbol;
    type 1 0;
    line c1 c2;
    overlay.

The 1 of the *type* 1 0 subcommand specified use of the circle symbol for the data, while to 0 of this subcommand specified that no symbol was to be used for the true line. The line c1 c2 subcommand connected the true line points (or curve points if those were what was generated in columns c1 and c2) and the overlay subcommand put all points on the same graph.

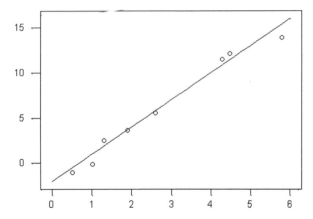

Figure 2-10   Data with Error with its Fitted Straight line

Repeat this graphing process with various different distributions chosen to produce the error (try some larger standard deviations as well as distributions like the uniform and Weibull) and formulas other than a straight line, to see the kind of data you get when error is added to various formulas. Seeing what you get with this variety of models should provide some feel as to what you might see in practical applications.

Summary

The purpose of this chapter was to provide an overview of what probability and statistics are all about, and to give an indication of the breadth of these subjects. The availability of sophisticated computer programs allows a real *hands-on* approach, making it possible to work with interesting and challenging problems almost from the start, and to find satisfying solutions quickly.   Following Chapter 3 we will fill in many of the details that have been glossed over, and begin a systematic development.

Problems 2.17

1. Suppose that the number of phone calls initiated in a given building in a 5 minute period has a Poisson distribution with mean 3 (i.e. an average of 3 phone calls in 5 minutes). Roughly what is the probability that the number of initiated calls is even?

   Hints: Compute the probabilities of 0 calls, 2 calls, 4 calls etc., by first using the Minitab *set* command to put, say, 1,2,...,30 in column 1. Then using the command let c2 = 2*c1 puts 0,2,4,... into column c2. Compute the probabilities for each element of c2... .

2. Determine how to use the uniform distribuion to generate any discrete distribution.

Review Exercises 2.16

1. What is meant by *statistical regularity* ?

2. What is a *macro*? What are the uses of macros?

3. How would you use the macro *meanseq.mac* to help you determine for which probability values statistical regularity takes hold faster?

4. What type of probability distribution is most common for describing *measurement errors*?

5. Do successive sample medians seem to exhibit statistical regularity? Justify your answer.

   Hint: It may help to modify the macro *meanseq.mac* to handle sample medians.

6. How are probabilities of intervals related to probability density functions?

7. What does a histogram of a large sample of data from some distribution often look like, and why?

8. What accounts for the discrepancies between a histogram and the probability density of the distribution from which the sample generating the histogram was drawn?

9. Suppose that a production process produces successive units each of which has a 3% chance of being defective. What is the probability that in 1000 such units the number of defectives is

   a. exactly 30

   b. between 25 andd 35 (careful here, because you want to include both 25 and 35. The cumulative probability of being <= [less than or equal to] x is what is computed by the binomial cumulative probability evaluated at x.)

10. Structurally, what distinguishes statistics from probability?

11. What is the basic objective of statistics? Illustrate with some examples.

12. What is a *t*-test used for? Why are t tests important?

13. What is the justification for the common assumption of a straight line relationship between two variables?

    What are the limitations of the straight line assumption?

    Why is so much emphasis placed on testing that $a = 0$ in the model $y = ax + b + error$?

Chapter 3

Sampling and Descriptive Statistics

3.1 Representative and Random Samples

Much of the data that we gather to gain information can be thought of as a sample from some larger collection of objects — which we will refer to as the *target population*.[1] When we sample from such a population we would like the sample to be *representative* for some attribute. By this we mean that we want the proportion of objects in the sample with this attribute be close to the proportion of the population[2] with this attribute — because it is usually the population proportion with the given attribute that is of interest. For instance, just recently we had good reason to believe that our dog (a not too discriminating greyhound) had eaten an entire container of Tums (this is the same dog that chewed a full tube of crazy glue). The veterinarian took a sample of his blood. Unless this sample's proportion of calcium (the main ingredient in Tums) was likely to be close to the proportion of calcium in his entire bloodstream, it might be difficult to decide on the proper treatment. Similarly, when a chef samples from a cauldron of soup, he hopes that the flavor of the sample is close to the flavor of any bowl that is served.

For reference purposes we supply the following formal definition.

Definition 3.1: Representative sample

A sample from some population is said to be **representative for some attribute** of the population (such as the proportion of blond people in the population) if the proportion of objects in the sample with this attribute is close to the population proportion (or probability, in the case of conceptual populations) with this attribute.

❑

Practice Exercises 3.1

1. What attribute of the population are we trying to determine from a presidential election poll?

2. What attribute of a patient's blood might we be trying to determine from an HIV test?

---

1. So a population need not necessarily refer to a collection of people or animals. It can be a conceptual collection, e.g., the collection of all possible voltages across some electronic device, or an actual collection such as the soup in some pot.
2. For some conceptual populations we may not be able to refer meaningfully to a *proportion* of the population — for instance, the population of all time intervals [0;1], [1;2], [2;3],... . For such populations the word *probability* turns out to be more suitable.

3. If light bulbs are examined after being burned for 800 hours, what attribute seems to be the one being sought?

4. What attribute(s) might an insurance company want to determine when

a. setting a rate for medical malpractice insurance for surgeons?

b. setting a rate for medical malpractice for dermatologists?

c. setting the rates for fire insurance?

Having established the desirability of representative samples in many situations, the obvious next step is to figure out how to achieve them. For soup and the bloodstream the answer is to mix well and take a large enough sample. To deal out playing cards for poker, pinochle or bridge, we shuffle sufficiently many times. There is a mathematical equivalent to the mixing and shuffling just mentioned — namely *sampling at random from a finite population.*

Definition 3.2: Random sampling from a finite population without replacement

Suppose that we have a population of n distinct objects, denoted by $x_1, x_2,...,x_n$. (Here, the population itself is denoted by $\{x_1, x_2,...,x_n\}$ and is a set with a finite number, n, of elements.)

We say that **we have sampled k (distinct) objects at random from this population if all different subsets consisting of k distinct objects from this population have the same probability of being chosen.**[1]

The sample so generated is called a **pure** or **simple random sample without replacement**.

☐

It can be shown (see Topic 4.22 on page 119) that a sample without replacement can be generated by drawing one observation at a time, but then disregarding the order in which the sample was drawn.

Problem 3.3

Many people have the mistaken notion that in order for a sample of size k without replacement to be a pure random sample, it is only necessary that each member of the population have the same chance of being chosen. Show that this is not the case.

Hint: Consider a population consisting of 50 men and 50 women if you first chose a person at random from the population, and then all remaining choices must have the same sex as the person chosen first?

Note that for sampling without replacement, we must have $k \leq n$. In sampling without replacement we are talking about **subsets, which are completely determined by their members.**

---

1. We will see what these probabilities are in Chapter 4 when the rules of probability are spelled out.

Another kind of sampling from a finite population is **sampling with replacement**.

Definition 3.4: Random sampling from a finite population with replacement

A sample of size k with replacement from the finite population $\{x_1, x_2, ..., x_n\}$. consists of a sequence of elements, $(y_1, y_2, ..., y_k)$, where each of these elements is one of the elements of the population $\{x_1, x_2, ..., x_n\}$.

The same element, say x_5, can occur many times in the sequence $(y_1, y_2, ..., y_k)$. Two samples with replacement, $(y_1, y_2, ..., y_k)$, $(z_1, z_2, ..., z_k)$ are considered **different** if there is at least one integer value, j, between 1 and k for which $y_j \neq z_j$ [1].

The sampling is called **random and of size k if all samples are of size k and all distinct such sequences are equally likely.**

◻

Because the object generated by sampling with replacement is a sequence, it would appear implicitly that the order in which the objects are drawn is being taken into account. It will be evident in Topic 4.22 on page 119, how this kind of sample can be generated by drawing one element of the sequence at a time. In either case (drawing one element of the sequence at a time, or drawing the entire sequence at once) the two samples with replacement $(1, 2, 3)$ and $(2, 1, 3)$ are different, while the two samples $\{1,2,3\}$ and $\{2,1,3\}$ without replacement aren't different. If 1,2, and 3 here represented football plays, we would want to describe what is done in three successive downs as a sample with replacement — because we should think of $(1, 2, 3)$ (first using play 1, then play 2 and finally play 3) as a different strategy from $(2, 1, 3)$.

If we sample a single object at random from a population of n distinct objects, all individuals have the same chance of being chosen, namely $1/n$. If we sample two individuals without replacement at random from this population, the probability of any specific individual sample turns out to be $2/[n(n-1)]$. The validity of these assertions follows from the rules of probability, to be introduced in Chapter 4.

Notice that it is conventional to think of sampling without replacement from some finite population as yielding a set of objects from this population, while sampling with replacement yields a sequence. We could have defined another type of sampling without replacement from a finite population, in which a sample is a sequence. As we will see, the samples yielded by Minitab are always sequences, but when we want a

---

1. You might ask why such a fuss is being made to distinguish sampling with replacement from sampling without replacement. One reason is that there are times when a person might really want to sample with replacement — for instance a football coach choosing his plays would surely want to reuse plays which seem to be working. If he had to sample without replacement from his set of possible plays, he would be at a big disadvantage. (It might not even be possible for him to use different plays at each down.) On the other hand, if destructive testing of objects from a production line is being done, it is impossible to sample with replacement.

sample without replacement from a finite population we only consider the set of elements in the sequence that Minitab generates (and not the order).

Assertion 3.5: The importance of random sampling

A random sample from some population is highly likely to be representative for any given attribute of this population, provided its size is sufficiently large.

☐

This result is self-evident for sampling without replacement; to prove it *generally* requires the machinery developed in Chapter 4.

Most people have little trouble believing that *the mean of the measurements on a large enough random sample with replacement is usually close to the mean of the measurements on the entire finite population that the sample came from.* But without experience with such samples, most of us don't have a clue as to how large a sample is needed to have a good chance of achieving a specified accuracy. The truth of this assertion is illustrated time after time by the unwillingness of so many people to believe that a random sample of 1112 voters can be used to determine the proportion of voters in a population of 200,000,000 adults in favor of a particular presidential candidate to within .03 roughly 95% of the time or better. Or by people who say *How can the polls be accurate, since I've never been polled in my life?*[1] Ask someone with no experience in probability what the probability of exactly 50 heads in 100 tosses of a fair coin, or what the probability of 60 or more heads in 100 such tosses, and the answers you get are almost always very far from the correct ones. To try to overcome these shortcomings, we will spend the rest of this section getting a feeling for the properties of random sampling by doing the next best thing to real experimentation — namely computer simulations of sampling from arbitrary finite populations, as well as from finite populations constructed by sampling from theoretically described conceptual populations. This will be followed by a comparison of various sample and population properties. We start by looking a sampling from a theoretically described population.

Topic 3.6: Sampling from a theoretical population

In Section 2.2 of Chapter 2 we got some experience in sampling from theoretical populations. The properties of the values that were generated were determined by the *interval probabilities* of the theoretical population. Usually an interval probability (probability that the outcome was in the given interval) was determined either as the area lying over the interval and under the probability density function, or as the sum of the probabilities of all points with nonzero probability in the interval. In the probability density function case, any generated sample (in theory) consists of

---

1. The main problem with obtaining reasonable accuracy in estimating the probability of some attribute from a small random sample is not its small size, but the failure to achieve the desired randomness. To help in constructing representative samples it is necessary to have a trustworthy way of generating random numbers, and to use some form of random sampling.

distinct values. In the situation described by a probability function, a generated sequence could have repeated values.

The following example illustrates how to obtain a sample of specified size from a given *normal* theoretical population.

Example 3.2: Sample of size *n* from a normal population

We want to put a sample of *n* independent[1] values from a (theoretical) normal population with mean μ and standard deviation σ into a specified column, e.g., column c1. We will do this by means of *line commands* typed in the Session window, rather than by mouse pointing and clicking. Recall what was done in Example 1.3 on page 10 to enable these line commands (First left click with the mouse in the Session window. Then from the top menu bar select Editor -> Enable Commands via the mouse). Then at the Minitab prompt (MTB >) type

MTB> random n c1; <Enter>

MTB> normal μ σ. <Enter>

All 3 symbols, n, μ, and σ stand for numerical values to be typed in, with *n* a positive integer and $\sigma \geq 0$. The prompt is supplied by the computer. <Enter> indicates that the Enter key is to be pressed. Note the terminating semicolon ; and period. The numbers being sought will appear in column c1 in the worksheet. ∎

Practice Exercises 3.3

1. Generate a sample of 100 values from the uniform distribution on [0; 1].

2. Generate a sample of 25 values from the Poisson distribution with mean .5. Do you observe repeated values?

3. Generate a sample of 2500 values from the Poisson distribution with mean 7. Use the describe command to determine the sample mean, and note how close it is to the theoretical mean value of 7 .

4. Generate a sample of 2500 values from the normal distribution with mean 7 and standard deviation 2. Use the describe command to determine the sample mean and sample standard deviation, noting how close they are to the theoretical values.

Topic 3.7: Random sampling from a finite population

Minitab is capable of obtaining random samples, both with and without replacement, from a finite population. It does this by sampling from the set of nonempty rows in a designated column, and putting the sample in some designated column (possibly the same as the one from which the sample is taken). If the column from which the sample is taken has distinct entries, then the column represents a well

---

1. By *independent* values, we mean that knowledge of any of the generated values does not change what we expect the remaining values to be. This concept will be clarified and made more precise in Definition 4.29 on page 127.

defined population, and the sample is a random one from this population. If there are repeated entries, with some appearing more than others, the sample will not be a random one from the population of distinct entries to this column; by sampling with replacement from a column using Minitab, the probability of obtaining a particular distinct entry is

$$\frac{\text{the total number of rows with this entry in the column}}{\text{the total number of nonempty rows in this column}}.$$

It is now time to illustrate how Minitab samples from a finite population.

Example 3.4: Random samples without and with replacement

Suppose we want to choose a random sample of size 20 without replacement from a production lot of 1000 units, numbered 1 to 1000. First we fill column c1 with the numbers 1 to 1000, via choosing Calc from the top menu bar, selecting Make Patterned Data from the drop-down menu, and choosing Simple Set of Numbers.

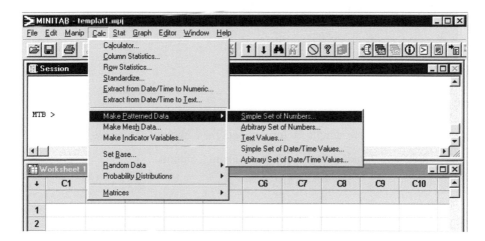

This results in the dialogue box below, which I filled in as shown.

Simple Set of Numbers		☒
	Store patterned data in:	c1
	From first value:	1
	To last value:	1000
	In steps of:	1
	List each value	1 times
	List the whole sequence	1 times
Select		
Help	OK	Cancel

Clicking on OK resulted in the corresponding command line in the session window and the *lot* numbers generated in the worksheet as illustrated below.

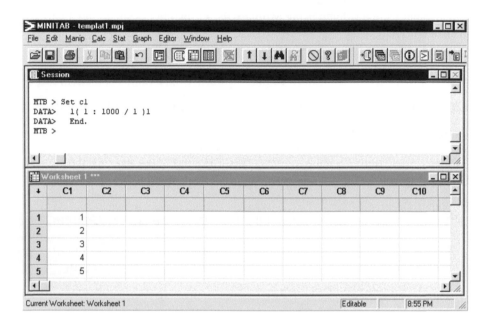

Now if you want to sample 20 lot numbers from column c1 you choose

Calc -> Random Data -> Sample From Columns...

from the top menu bar as shown below

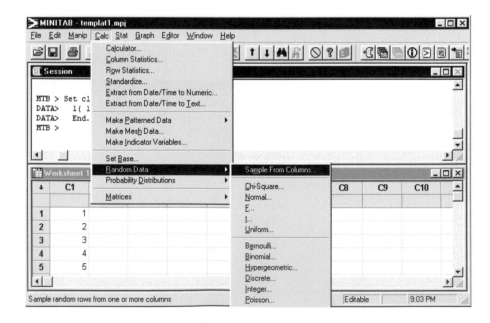

to obtain the following dialogue box, which I filled out as indicated.

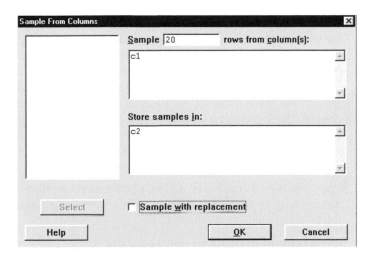

Clicking OK yielded the output in column c2 below and the last command in the session window.

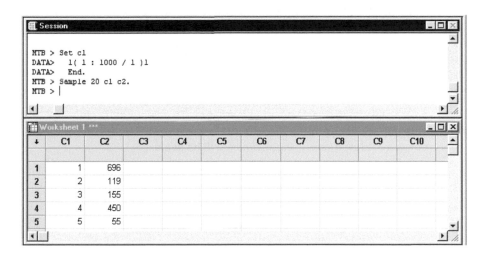

Note this last command line in the session window, which shows how to get the sample of 20 lot numbers from column c1, without replacement and put them in column c2. If we had wanted to sample with replacement, the command would have been

MTB > Sample 20 c1 c2;

MTB > Replace.

Problems 3.8

1. Suppose that in a population of 1000 people there are 300 smokers (numbered 1 to 300) and half the population is female (the females being the even numbered people). Choose a random sample of 250 from this population without replacement, and observe what proportion of the people in the sample are smokers, and what proportion are female. Does your sample seem representative for gender and smoking behavior?
   Hints: Put the numbers 1 to 1000 in column c1. Put the sample in column c2. Sort column c2 and put the results in column c3. Now it is easy to count the number of elements of the sample less than or equal to 300. To count the number of even entries in the sample, divide the elements of column c3 by 2 and put this is column c4.

2. Figure out how to randomly divide a population of 900 objects into 3 nonoverlapping equal size subpopulations. Why might you want to do this?

Example 3.5: Mean of sample measurements

We will create a finite population of distinct values by sampling from a theoretical population having a specified probability density. We just reviewed how to do this from a normally distributed population in Example 3.2 on page 51. Here, after enabling the Minitab command prompt (click in the Session window and from the Menu bar at the top chose Editor -> Enable Commands) type

random 300 c2; <Enter>

normal 2 .75. <Enter

and you will see the values of 300 independent normal variables with population mean value 2 and population standard deviation of .75 entered into column c2 of the worksheet. Because the normal distribution has a probability density, probability theory shows that almost surely, these values will be distinct.

Next we will choose 400 values at random with replacement from the population just created and put them in column c1. As shown in Example 3.4 on page 52, we do this by the session window command lines

sample 400 c2 c1; <Enter>

Replace. <Enter>

The easiest way to compare the most important of the simple properties of the entries in columns c1 and c2 is the command

Describe c1-c2 <Enter>

The result you get, will of course depend on the values in columns c1 and c2. Here, preceded by the Session window's command line version of what was done with the *point-and-click* commands, is what resulted in the run carried out when this chapter was written.

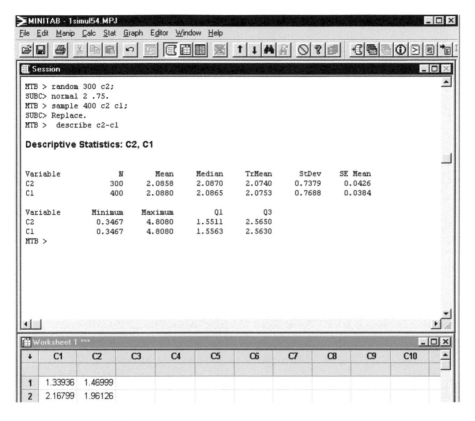

The only matter of interest here is that the various descriptive statistics from the sample of size 400 in column c1 are close to the population values from column c2. It is instructive to run the macro *meanseq.mac* on the data in column 1, to see how the successive averages approach the final mean of 2.042 printed above.

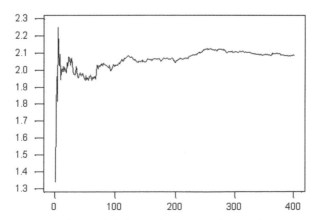

Figure 3-1 Successive means of a sample

Prior to running this macro, we had to indicate how many data points are used — in this case we typed   let k2 = 400 <Enter>, to obtain Figure 3-1.

If we set k9 equal to 20, by the command   let k9 = 20, and use the version of meanseq which uncomments the three commands preceding the last two, we obtain the following expanded version of the above (omitting the first 20 points) shown in Figure 3-2.

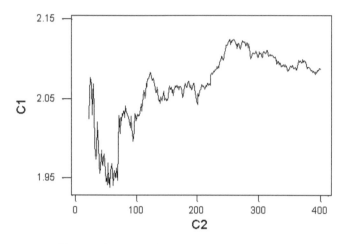

Figure 3-2 Expanded scale, omitting first 20 means

The sample means seem to be settling down, but they are doing it in a rather jagged manner. Thus, while the successive sample means *seem* to be approaching the mean of the population from which they came, and generally the larger the sample, the more certain we can be that the mean of the sample will be close to the population mean,[1] we cannot claim that the sample means are closer to the population mean when the sample size is larger. We will see (Problem 3.10.2exernum) that the means of larger samples can (in a certain sizeable proportion of these samples) be further from the population mean than many of the means from the smaller samples.

■

Problems 3.9

1. In a hospital two different treatments are to be tested on the 150 patients
   with hepatitis C. Any patient can receive only one of these treatments.
   How would you choose the patients to receive treatments 1 and 2.
   (Justify and carry out your choice.)
   Hints: Better decide whether to sample with or without replacement.

---

1. In particular if we know a specific finite length interval which includes all
   possible data, and we are given an allowable nonzero error, for any specified
   probability less than 1 we can determine a sample size guaranteeing that with at
   least this probability our estimate of the mean based on this sample size is no
   further from the population mean than this allowable error (see Chapter 5).

Then, just *number* the people in the hospital population. Also look at Calc -> Make Patterned Data from the menu bar.

2. On page 50 it was claimed that *a random sample of 1112 voters can be used to determine the proportion of voters in a population of 200,000,000 adults in favor of a particular presidential candidate to within .03 roughly 95% of the time or better.* Assuming that the voters sampled tell the truth, how would you convince yourself of the truth of this assertion?

    Hints: The 200,000,000 figure is a red herring. Any population size could be used, because under sampling with replacement any one observation has probability p of declaring for the candidate favored by a proportion p of the voting population. So you might just as well sample from a theoretical population having two values, 1 with probability p and 0 with probability $1 - p$ — i.e., sampling from a Bernoulli population.

    Next you might ask *which is the **worst case** value of p?*   i.e. the one which would require the most data? A reasonable answer is that it is the value of p which yields the least predictable answer. Knowing that $p = 0$ and $p = 1$ yield perfectly predictable results should make the choice of *worst case p* fairly obvious.

    You can then use
    > the *rmean* command (or Calc -> row statistics)
    > and the *sort* command.

    (If you get a proportion equal to .5 [very unlikely], in Florida you may assign the vote to the Republican, in Illinois to the Democrat.)

3. Run meanseq.mac on a sample of size 1000 with replacement from a population of size 100 values generated from the standard normal distribution to see how the successive averages settle down to near the mean of the population. (It may take quite a while — like 20 minutes on some older machines — for the graph to be generated.)

We conclude this section with an optional look at how tables of random numbers were used to choose samples in the era before everyone had access to PCs.

It is worth mentioning that the random numbers generated on a computer are actually produced by a deterministic algorithm — which, however produce numbers which seem to have many of the properties supposedly possessed by random numbers. As a precaution, it is always safest to try such *pseudo-random* numbers on a problem similar to the one you are working on, whose answer you know. If the answer you get is not close enough[1] to the known correct answer, then you would be well advised to avoid the random number generator that produced them.

To obtain a random number table of the type that was used before 1970, here is a

---

1. Just how close *acceptable* is will be discussed in Chapter 5 for some important cases.

simple macro, *rnum.mac*:

GMACRO

rnum

random 10 c1-c10;

unif 0 1.

do k1 = 1:10

let ck1 = round(10000*ck1)

enddo

ENDMACRO

Running this macro produces the following 10 by 10 table,

↓	C1	C2	C3	C4	C5	C6	C7	C8	C9	C10
1	66049	66007	18954	44018	51229	37891	10864	23274	91196	75180
2	78972	47607	25079	51039	44588	69401	83104	15080	28244	24259
3	03051	08544	70580	55173	8390	5086	76491	76821	99910	9651
4	62531	73714	20903	92248	14899	84281	23691	50512	34844	3148
5	19812	68752	73460	16411	86626	75453	15371	43095	73634	53477
6	48828	35044	81816	41324	9991	68564	37211	13539	35503	70149
7	64052	66770	14734	9746	10144	62495	56329	10386	35576	51787
8	24510	4038	35633	50664	87498	86493	43647	88105	79321	84819
9	90334	99195	77964	35298	43986	62221	61296	58427	50095	18016
10	92158	54227	13863	61840	42896	60502	26680	65231	21856	231
11										

which is *almost* what we want. The leading 0s are missing from some entries, so you need to fill them in by hand (which I have done for two of them; there are 11 of them all told).

Example 3.6: Choosing 2 distinct people from a population of size 5

We start out by numbering the people in the population from 1 to 5. We can then go through successive digits in the table — across rows or down columns. The important point is to never use any digit more than once. Suppose we go across rows, starting at the left side of row 1. Our first 10 digits are

$$6 6 0 4 9 6 6 0 0 7$$

(just disregard extra spaces). Since we only need to choose from 5 people, disregard all digits outside the range from 1 to 5. So we first choose person 4. We now go on to the next 10 digits which are

$$1 8 9 5 4 4 4 0 1 8$$

We see that the next person is number 1 and the task is done, with the choice of persons 4 and 1.

Note that if we had to choose 2 people from among 50, we would need to use two digits at a time, disregarding 0 0 and any pair above 5 0. In this case we would have chosen person 04 and 07 (persons 4 and 7). ■

Practice Exercises 3.7

1. Use the table provided above to choose a random sample of 4 distinct people from a population of size 1000.

2. Use the table provided above to choose at random groups of size 2, 975 and 3 distinct people from a population of size 980.

3. Use the table provided above to choose at random groups of size 2, 1001, and 3 from a population of size 1006.

L

3.2 Descriptive Statistics of Location

Descriptive statistics is a catch-all term for functions of observed data which reflect some population property. When all you have available about some population is observable data, examination of the more important descriptive statistics is a vital first step in getting a feel for the population's behavior.

It isn't difficult to clarify what is meant by *descriptive statistics* when the parent population is a finite one. So, for example, the mean of a sample drawn from a finite numerical population (one with a finite number of numerical elements) is a descriptive statistic, since the sample's mean (as we have seen earlier) tends to be close to the population's mean if the sample size is large enough. The sample mean is, in a sense, a measure of a finite population's location because:

if every population element is shifted by an amount a, then both the population mean and the sample's mean are shifted by this amount;

these population (sample) mean lies between the smallest and largest elements of the population[1] (sample).

The *median* of a population is also a quantity which is a measure of the population's location. The median of a population is essentially the *middle* value. More precisely, if *med* denotes the median salary of some population of people, and s_i denotes the salary of person i from this population, then at least half of the people have a salary $\leq$ *med* and at least half of the people have a salary $\geq$ *med*. So the median is the 50-th percentile. There may be lots of medians, say if the respective salaries of four people were $20,000, $20,000, $40,000, $50,000 then any number in the range from $20,000 to $40,000 would be a median of this population of salaries. In Minitab, when this situation occurs, the midpoint of this interval is displayed as **the** median. The sample median (median of a sample from a population) is defined as above — always including any duplications in determining the *middle value.*

Although both the mean and the median are measures of location, they often reflect quite different population properties. The mean is strongly influenced by individual values with extremely large magnitudes. So in a small city, just one *Bill Gates* would affect the mean income, but would have very little effect on the median income. When the median and mean are quite different, the median would seem to give more information about individuals in the population. So, as a measure of how well the population is doing, the population median would seem to be more

---

1. It isn't hard to convince yourself of the truth of this claim, either by examining some specific cases, or using elementary algebra.

informative. However, for the purposes of the Internal Revenue Service, the mean income might be more useful.

When the target population from which we draw a sample is not finite, at this stage of our development what is meant by *descriptive statistics* is not clear, because we have not yet defined the mean or the median (or most other properties) of such a population. What is missing, and will be supplied in Chapter 4, are the definitions of population means, medians, standard deviations, etc., for populations described by discrete probability functions or probability density functions. Suffice it to say here, that with the aid of the frequency interpretation (Definition 2.1 on page 19) *these definitions will be introduced as the quantities related to the discrete probability function or the probability density that the corresponding sample descriptive statistics get close to as the sample sizes get large.*

There is no guarantee that a given sample mean (or sample median) is farther from the population mean than a sample mean (or sample median) based on a larger sample size. The easiest way to see this is to look at samples of sizes 2 and 3 with replacement from the population consisting of the numbers 0 and 1; because — the mean of the population is 1/2, but a sample of size 2 can have a (sample) mean (or median) of 1/2, while a sample of size 3 cannot.

When the population median is not unique, the behavior of the sample median (at least as computed by Minitab) tends to bounce around, never really settling down; we illustrate this phenomenon in Problem 3.10.1. below. When a finite population (or a population described by a probability density) has a unique median[1] then it can be guaranteed that for sufficiently large samples with replacement (or repeated samples from the density) the sample median is likely to be close to the population median (although it may take quite a large sample before this takes hold, even if all elements of the population are known to be between two specified values. You can see why this happens in Problem 3.10.2. to follow).

Problems 3.10

1. Generate 20 samples of sizes 100 and 200 with replacement from the population consisting of 0 and 1, and determine the medians of these samples. Explain the behavior that you observe.
    Hints: We recommend that you use Calc -> Random Data -> Integer with each sample size in the *Number of Rows* slot, and store the results in columns c1-c20. The minimum value is 0 and the maximum is 1. When you have generated the results, it is easiest to type the command describe c1-c20 to determine the separate medians, even though this is overkill. (Sampling from the column consisting of 0 and 1 is not suitable, since it generates 20 identical samples.)

2. Generate 10 samples of sizes 20, 1000, and 10,000 from the discrete distribution with values 0, 1 ,2 and respective probabilities .49, .02, .49.

---

1. To satisfy curiosity about medians of populations defined by probability densities, the probability of a value $> (<)$ a median equals 1/2.

For each sample size compute the medians of these 10 samples, and see
if you can explain the behavior you observe.

Hints: i. You may want to recall Problem 2.2exernum.

ii. Think about what happens to the proportions of observed
0s, 1s and 2s as the sample size gets large.

3. Estimate all of probabilities for the sum on the toss of two fair dice,
giving some thought for how large a sample you need.

   Hint: Save trouble by using density Histograms — but be careful that
   your answers make sense (remember that the probabilities should sum to
   1, because of their relation to proportions).

4. Estimate the proportion of times that the mean of a standard normal
random sample of size 20 is closer to the population mean (which is 0)
than the mean of a sample of size 40.

   Hints: The row-means command, *rmean* makes this problem fairly
   easy. Generate, say, 20 standard normal random samples of size 100,
   putting them in columns c1-c20. To do this, after the Minitab command
   prompt MTB> type rmean c1-c20 c41. The sample means of the 100
   rows will be put into column c41. Next, generate 40 standard normal
   random samples of size 100, putting them in columns c1-c40 and put
   the row means of these samples into column c42.   To answer the
   proposed question you want to see what proportion of elements of
   column c41 is closer to 0 than the corresponding elements of column
   c42. You can just look at columns c41 and c42 to get the answer. If you
   want to see how this can be done completely by the computer, you can
   read the advanced material in small print which follows.

   It seems easier to phrase this question in terms of the squares of the
   elements of columns c41 and c42 — since x is closer to 0 than y precisely
   when $x^2 < y^2$. (You need to create two new columns for the squared
   values, via let c43 = c41*c41 and let c44 = c42*c42.) If you now type
   let c45 = c44 - c43, column c45 will have a negative entry precisely when
   the element of c43 is larger than the corresponding element of c44. Figure
   out how to convert the negative elements of column c45 to -1 and the
   positive elements of c45 to 1 (refer back to Example 1.3, in particular the
   discussion beginning on page 13). Then figure out how to convert the -1
   to 0 and the +1 to 2, and finally (without changing the 0) convert the 2 to
   1. Then the mean of the column just generated is the proportion of 1s,
   which is the proportion of pairs in which the sample size 20 mean is closer
   to 0 than the sample size 40 mean.

To help get more comfortable with statistical data, we will graph a random sample
of size 400 from a standard normal distribution, which will provide an idea of the
values (the y coordinates of points on the plotted graph) as well as the variability to
be expected in this case. We will do this by means of the time series plot command,

tsplTot (which can be considered a part of our arsenal of descriptive statistics tools). Before running this command, in Minitab we put the sample of size 400 in some column, e.g. column c2 (via the random data command, see page 11 using the mouse, or via typing   random 400 c2   <Enter>).

Now just type
tsplot c2; <Enter>
connect.   <Enter>

and you will see a graph something like the following.

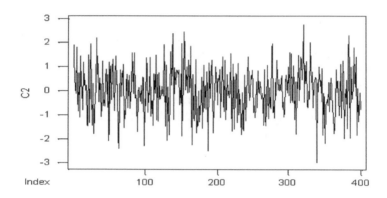

Figure 3-3   Successive Standard Normal Data Values

Topic 3.11: Long-run usual (and unusual) behavior of successive sample means: $x_1, \frac{1}{2}(x_1+x_2), \frac{1}{3}(x_1+x_2+x_3),...,\frac{1}{n}(x_1+x_2+...+x_n)$

We look at plots of successive means when plotted against their positions. The macro *meanseq.mac,* which we introduced on page 20 does this plotting for us. In order to run this macro you first decide on the sample size, and in Minitab store this value in k2.

If you want a sample size of 200, you type

let k2 =200 <Enter>.

Then if you want 200 standard normal variables, type

random 200 c1 <Enter>.

If this macro is stored in the file c:\meanseq.mac, then to run it from Minitab you need only type

%c:\meanseq <Enter>.

You will get a graph looking something like the one in Figure 3-4 which follows.

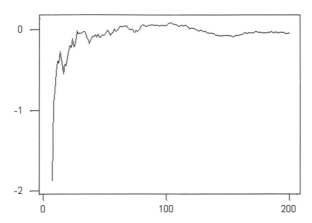

Figure 3-4 Successive Normal Means

Notice the sort of ragged approach of the sequence of sample means to the population mean (of 0). The macro, *meanseq.mac* can be used to investigate how sample means of random samples from any probability density behave. Below are some exercises to see how useful this macro can be. No claim is made for the efficiency of the code here. Samples larger than 300 may take quite a while to run on today's machines (year 2001).

In some of the problems to follow you will notice that **there are some populations** described by probability densities (of which the Cauchy is the most well known), **in which successive means do not ever settle down** (which indicates that there is no population mean). This phenomenon occurs in populations where values of extremely large magnitude occur with too great a frequency.

Problems 3.12

1. Run the macro *meanseq.mac* on a sample of size 200 from a normal density with mean 0 and standard deviations of 2, 4, and 8. What does this tell you about the meaning of the standard deviation?

2. Use the *tsplot* command on Poisson random samples of size 100 and means 10, 30, and 100. What seems to be happening to the variability of the data as the mean increases?

3. Run the macro *meanseq.mac* on the random samples of the previous problem. What seems to be happening to the variability of the successive sample means? Run the *tsplot command* and the macro *meanseq.mac* on Cauchy samples of size 200. Does it look as if the sample means are settling down as the sample size gets bigger? Explain the behavior you observe. (You might want to look at the graphs done by *plotsamp.mac* and *meanseq.mac* for Problem 4. in this set, of the same Cauchy sample of size 400.)

4. Run the *tsplot* command and the macro *meanseq.mac* on the Bernoulli distributions with probability of success .52 and sample size of 300. Does it look as if a sample size of 300 is adequate to determine from looking at this data to determine whether or not to predict the popular vote winner in a presidential election when the population proportion of voters for a given candidate is .52 or larger? Fill the first 1000 rows of columns c1 to c300 with random samples from the Bernoulli distribution and see if you can use this data to try for a more convincing answer to the previous question. *Hint*: Compute the row means (using the *rmean* command — see page 22). Run the *tsplot* command on Cauchy random samples of sizes 5, 10, 50, 200. What does this seem to indicate about the nature of the Cauchy distribution, as compared to the standard normal distribution. Below in Figure 3-5 is a graph of the output of the *tsplot* command, run on a Cauchy sample of size 400.

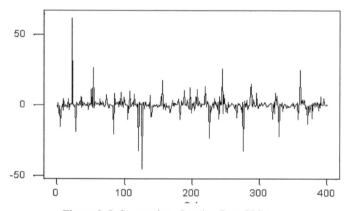

Figure 3-5 Successive Cauchy Data Values

5. Run the macro *meanseq.mac* on a Cauchy random sample of size 300. What does the graph you see indicate about whether the successive means of Cauchy variables ever settles down? Below in Figure 3-6 is a graph of a Cauchy sample of size 400 generated by the macro *meanseq*.

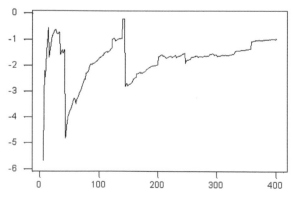

Figure 3-6 Successive Cauchy means

⌈

6. Modify the macro *meanseq.mac* (page 20) to compute successive *medians*
    instead of successive means. Calling this macro *medseq.mac*, apply it to any
    of the previous problems where *meanseq* was applied, comparing the results to
    the *meanseq* results. The results you observe should convince you that for the
    Cauchy distribution, the sample median is a much better estimate of location
    than the mean.

⌊

3.3 Descriptive Statistics of Variability

The presence of variability that is not explicitly accounted for, is what
distinguishes statistical from deterministic models. In order to come up with
statistical models whose accuracy is known, we must be able to get a handle on the
variability of such models.

A number of descriptive statistics appear reasonable as measures of variability.
The first one that comes to mind might well be the *range*, which is the difference

largest value − smallest value

in the sample. This descriptive statistic is the best for the uniform distribution, but
for most distributions that we will encounter, it is very poor; in particular, it does not
settle down for any normal distribution (becoming arbitrarily large as the sample size
grows large, but very slowly).

A more stable measure of variability is the *interquartile range*, which is the
distance from the 25-th percentile of the sample to its 75-th percentile.[1] Importantly,
if the units of measurement of the sample changed, say from inches to feet, the real
meaning of any measure of variability should be the same. If the sample's elements
were measured in feet, it seems reasonable to require that the measure of variability
should be in feet (as opposed to square feet, which would not be an easily
interpretable measure of variability of the quantity being measured). Furthermore, if
each member of the sample is shifted by the same amount, the measure of variability
should remain the same. The interquartile range satisfies these requirements.

Another measure of variability favored by some because it is less sensitive to
values of large magnitude than the most commonly used such measure (the standard
deviation) is the *(sample) mean deviation,* whose formula is given by

$$\frac{1}{n}[|x_1 - \bar{x}| + |x_2 - \bar{x}| + \ldots + |x_n - \bar{x}|]. \text{ For the definition of } \bar{x} \text{ see below.} \quad 3.1$$

We will not devote any substantial space to this particular measure, going on to
concentrate on the **standard deviation** — which, for reasons of mathematical

---

1. Any other similar distance, the k-th percentile to the $(100 - k)$-th percentile would
    serve similarly, though maybe less efficiently.

tractability (that are by no means obvious without substantial investigation) is the most commonly used measure of variability.

Definition 3.13: Population and sample standard deviations

Let $\{x_1, x_2,...,x_n\}$ be a finite population consisting of n numerical elements, and let $\bar{x}$ denote the population mean —

$$\bar{x} = \frac{1}{n}[x_1 + x_2 + ... + x_n] \qquad\qquad 3.2$$

and let

$$s = \sqrt{\frac{1}{n}[(x_1 - \bar{x})^2 + (x_2 - \bar{x})^2 + ... + (x_n - \bar{x})^2]}. \qquad\qquad 3.3$$

The value s is called the *population standard deviation*.

If $(y_1, y_2,...,y_k)$ is a random sample of size k **with replacement** from the population $\{x_1, x_2,...,x_n\}$ (Definition 3.4 on page 49) the *sample mean*, $\bar{y}_k$, is given by

$$\bar{y}_k = \frac{1}{k}[y_1 + y_2 + ... + y_k] \qquad\qquad 3.4$$

and the **sample** standard deviation, s_k is defined as

$$s_k = \sqrt{\frac{1}{k-1}[(y_1 - \bar{y}_k)^2 + (y_2 - \bar{y}_k)^2 + ... + (y_k - \bar{y}_k)^2]} \qquad\qquad 3.5$$

The reason for using k - 1 rather than k in the definition of *sample standard deviation*, stems from our desire to compute the sample standard deviation solely from available measurements. Thus, we could not use the population mean in this definition, but had to replace it by the available sample mean, $\bar{y}_k$. In the most commonly used models (the normal ones) this turns out to reduce the *effective sample size* by 1 observation (another non-obvious result).

❑

We can see that if the data displays no variability at all, the standard deviation is 0. Note also that if the scale of measurements is changed, say from feet to inches, the standard deviation increases by a factor of 12. Finally, it should be evident that the standard deviation does reflect the variability of the data about the mean, since the further any single datum is from the sample mean, the greater its contribution to the magnitude of the standard deviation.

The operational significance of the standard deviation arises from two major results, to be carefully introduced in Chapter 4 — namely the *Chebychev inequality* and the *Central Limit Theorem*. Both of these results concern the probability distribution of the distance (in standard deviation units) of an observation from the population mean. The Central Limit Theorem forms the basis for the *two-sigma rule-of-thumb* which will soon be introduced.

Prior to the widespread use of computers, a great deal of effort was put into hand computation of standard deviations. This is no longer necessary, thereby permitting us to put all of our emphasis on the uses of this measure of variability.

There are many situations in which the variability observed in our data can be thought of as a sum of quantities arising from a variety of similar sources. To mention a few:

- the number of people in a random sample of size $k = 200$ who plan to vote Democratic in a presidential election is the sum of 200 such quantities, where the i-th such quantity is the number of Democratic presidential votes (1 or 0) of the i-th person,

- the maximum blood pumping capacity of the heart of a randomly selected person from a homogeneous population of healthy adults might be considered a sum of the contributions of a number of similar genetic factors,

- the total roundoff error in adding up the results of repeated identical experiments.

If the number of elements constituting the sums referred to above is sufficiently large, the Central Limit Theorem asserts that under mild restrictions, the distribution of such a sum is approximately normal. Since roughly 95% of a normal population lies within two (population) standard deviations of the (population mean) it follows that in the previous examples, approximately 95% of such data (which can be thought of as sums) lies within two (population) standard deviations of the (population) mean.

Topic 3.14: The two-sigma rule-of-thumb

If each observation from a randomly chosen set of observations from some population can be represented as a sum of similar independent quantities, then, provided the number of terms in this sum is sufficiently large, the observations often have a distribution which is approximately normal. Since the probability that a normal observation will be within 2 (population) standard deviations of the (population) mean, roughly 95% of such data will fall within 2 standard deviations of the mean.

Example 3.8: Democratic percentage in a sample

Suppose that the proportion of Democratic voters in a presidential campaign is $p = .54$. It can be shown that the (population) standard deviation of the sample mean of a random sample of size $k = 1000$ with replacement from the voting population has a standard deviation of

$$\sigma = \sqrt{p(1-p)/k} = \sqrt{0.54(1-0.54)/1000} = 0.0157 .$$

So roughly 95% of samples of size 1000 will have Democratic proportions in the range from $.54 - 2 \times .0158$ to $.54 + 2 \times .0158$, i.e., $.5084$ to $.5716$, provided that the *two-sigma rule* is applicable. Turns out that it is, as can be verified using the binomial distribution; precisely, using computations similar to those in Example 2.2 on page 26, we find that when the proportion of Democrats in the population is .54, the probability is about .954 that the proportion of Democrats in a sample of size 1000 falls in the interval from .5084 to .5716.

■

Practice Exercises 3.9

1. Suppose the random numerical quantities $X_1, X_2,...$ are uniformly distributed on the interval from -.5 to .5. You can verify empirically that their standard deviation is approximately .29

   a. What proportion of **these** variables are within 2σ of the mean?

   b. Approximately what is the standard deviation of $X_1 + X_2$?
   Hint: Generate about 1000 samples of X_1, X_2, from which you can find 1000 repeated observations on $X_1 + X_2$. Then maybe use the Minitab describe command.

   c. Roughly what proportion of values of $X_1 + X_2$ are within two of their standard deviations of their mean, 0?
   Hint: The *sort* command might help here.

   d. Repeat parts b. and c. for $X_1 + X_2 + \cdots + X_{10}$.

   e. What conclusions do you draw from your computations?

2. The probability distributions of the number of cases of rare non-communicable diseases in some specified region are often described by Poisson distributions (see page 31). The mean value of the Poisson distribution used to describe the disease prevalence is the population size times the overall population mean number, λ, of cases per person. (This is often phrased in the form of mean number of cases per 100,000 people. So suppose for some particular dangerous disease, we accurately know λ to be close to 5.5 per 100,000 people — implying that the mean number of cases per person is .000055.) If you happened to find out that your city of size one million had 94 cases of this disease, might you be justified in worrying?
   Hint: First estimate what the standard deviation of the number of cases of this disease would be if this city was a representative sample from the entire population (by sampling from a Poisson distribution with mean value 55).

3. Suppose that when the process of producing inexpensive timers is in control (operating properly) the probability of a defective is .0062. A randomly chosen lot of 100 timers is found to have 5 defective ones. Does this seem to be trustworthy evidence that the manufacturing process is out of control?
   Hint: You want to estimate the probability of at least 5 defective units in a lot of 100 when the manufacturing process is in control. Check your result by finding the theoretical probability of this occurring (using the binomial distribution).

3.4 Other Descriptive Statistics

One of the most frequently used descriptive statistics, the *histogram*, was introduced on Example 2.3 on page 28. The area of the density histogram lying above any given interval should be reasonably close to the probability that an observation falls in this interval (provided sufficiently many observations were used in the histogram's construction). So a density histogram can provide a good estimate of the probability density of the data used in its construction. A *box plot* can be thought of as the *poor man's* histogram (in the sense that it provides a simple way of providing some of the same information as a histogram).   It is particularly easy to plot several box plots side-by-side, for comparing populations. The definition of a box plot given in Minitab is the following.

By default, a boxplot consists of a box, whiskers, and outliers.

outliers

whisker extends to this adjacent value—the highest value within upper limit

third quartile (Q3)

median

whisker extends to this adjacent value—the lowest value within lower limit

first quartile (Q1)

A line is drawn across the box at the median. By default, the bottom of the box is at the first quartile (Q1), and the top is at the third quartile (Q3) value. The whiskers are the lines that extend from the top and bottom of the box to the adjacent values. The adjacent values are the lowest and highest observations that are still inside the region defined by the following limits:

Lower Limit: Q1 $-1.5\,(Q3-Q1)$

Upper Limit: Q3 $+1.5\,(Q3-Q1)$

Outliers are points outside of the lower and upper limits and are plotted with asterisks (*).

To see how a box plot works, we first generate a few columns of data of different varieties using the five lines which follow, to yield the results in the worksheet below.

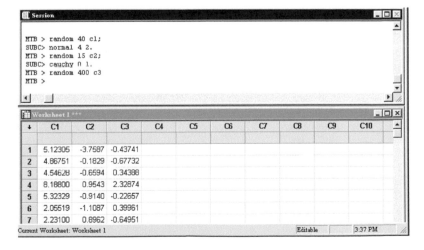

```
MTB > random 40 c1;
SUBC> normal 4 2.
MTB > random 15 c2;
SUBC> cauchy 0 1.
MTB > random 400 c3
MTB >
```

↓	C1	C2	C3	C4	C5	C6	C7	C8	C9	C10
1	5.12305	-3.7587	-0.43741							
2	4.86751	-0.1829	-0.67732							
3	4.54628	-0.6594	0.34388							
4	8.18800	0.9543	2.32874							
5	5.32329	-0.9140	-0.22657							
6	2.05519	-1.1087	0.39961							
7	2.23100	0.8962	-0.64951							

Current Worksheet: Worksheet 1   Editable   3:37 PM

Next, in order to get box plots for these three columns printed side-by-side, you are best off choosing the following from the menu bar,

Manip -> Stack -> Stack Columns

to obtain the following dialog box, to be filled out as shown.

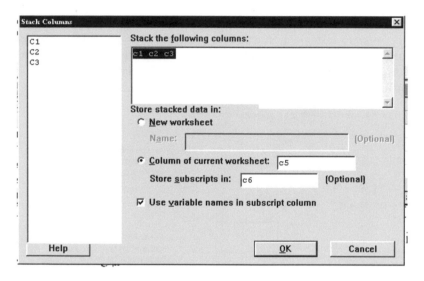

The corresponding command lines and effect on the worksheet below (after clicking OK) are shown next.

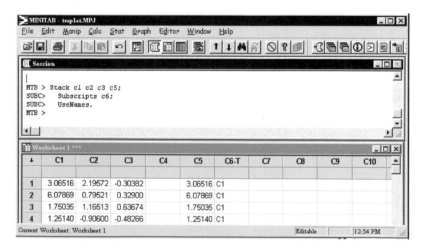

If you now type

boxplot c5*c6 <Enter>

in the session window following the MTB> command prompt, in Figure 3-7 you obtain the side-by-side boxplots that were desired.

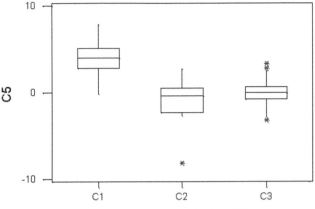

Figure 3-7   Side-by-side Box Plots

These box plots make it appear that the data from column c1 represent a population centered at a different location than those from columns c2 and c3. But we need a more solid foundation in order to be able to draw such conclusions reliably. This foundation will be developed in Chapters 4-6.

Practice Exercises 3.10

Generate data from a variety of distributions, and do side-by-side box plots, to get a feel for what you are likely to see under various different situations.

Problems 3.15

1. Construct 3 side-by-side box plots 150   observations from each population, the first being normal with mean 0 and standard deviation 1, the second normal mean 5 and standard deviation 1, and the third Cauchy with parameters 0 and 1.   What is the cause of the lack of information from these box plots?

2. How might you modify your data sets to make the box plots more informative?

Topic 3.16: Time series plots

In many situations we obtain data at equally spaced time points. To mention a few such cases:

- Medical monitoring of blood pressure in hospital intensive care units
- Seismograph records to detect earthquakes
- Measurements made on successive units from a production line, to monitor for drift from the desired specifications
- Measurements to monitor the economic health of the nation.

The natural descriptive statistic for such data is the *time series plot*.

The two graphsFigure 3-8 and Figure 3-9 to follow, were created by the sets of Minitab commands shown below.

```
MTB > random 100 cl
MTB > set c2
DATA> 1:11/.1
DATA> end
MTB > random 101 c3
MTB > let c4 = c2*c3
MTB > stack cl c4 c5
MTB > tsplot c5
```
TSPlot C5
```
MTB > tsplot c5;
SUBC> connect.
```

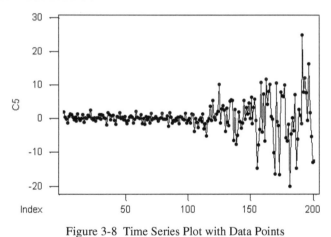

Figure 3-8 Time Series Plot with Data Points

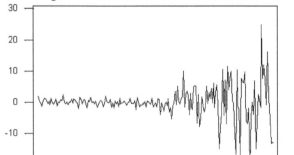

Figure 3-9 Unadorned Time Series Plot

In either of these plots, approximately where the time series underwent a change in character seems pretty evident. Often we want to know more — for example, what kind of change occurred, how big a change, etc. If these graphs represented, say, figures on volatility, we may need such information to determine what actions to take to reduce the volatility to acceptable levels. Determining such information requires some precise modeling, which must be based on a solid mathematical foundation.

Topic 3.17: Scatter plots

Ordered pairs, (x,y), of real numbers arise in many situations — sometimes as data pairs on individuals from some population (maybe   (LDL or HDL level, likelihood of a heart attack)[1] or maybe (log(SBP), log(P[MI in 5 years]))[2] in people) and sometimes as pairs whose first coordinate is a quantity controlled by the experimenter (such as the dosage of a drug like Lovastatin whose purpose is the regulation of the body's lipoproteins) and whose second coordinate is the response of the body to the given experimenter controlled quantity.   In both cases it is usually of interest to get an idea of the relation between x and y.   To get a handle on modeling the relation of interest, it seems natural to plot points (x_i, y_i) obtained from measurements on some set of animals or people. (Just imagine how difficult it would be to get any idea of such a relationship solely by staring at a stream of numbers.)

If a plot of such a set of pairs looked as follows,

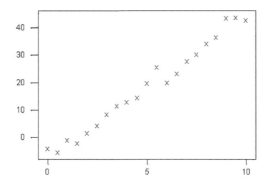

Figure 3-10 Data Appropriate to Straight Line Fit

then a straight line might be a reasonable type of curve to fit to this data.

---

1. LDL — low density lipoproteins, the *bad cholesterol*, HDL — high density lipoproteins, the *good cholesterol*.
2. SBP stands for *Systolic Blood Pressure*, MI stands for *myocardial infarction* (heart attack), *log* is the natural logarithm function.

If, on the other hand, the data looked as follows,[1]

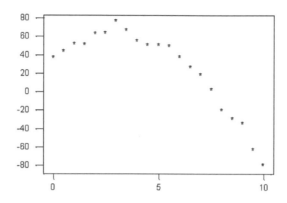

Figure 3-11 Data Appropriate to 2-nd Degree Polynomial Fit

no straight line seems appropriate to represent the data. A second degree polynomial fits the data much better.

In problems involving the effects of radiation, from theoretical reasoning, it's often reasonable to model data of the type in Figure 3-12 below as a weighted sum of negative exponential functions, $Ae^{-at} + Be^{-bt}$.

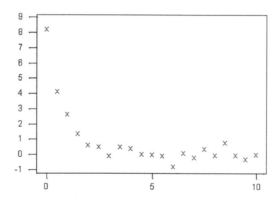

Figure 3-12 Amount of Undamaged Tissue vs. Time

Estimation of the values A,a,B,b requires sophisticated nonlinear regression programs.

---

1. The code to plot the data shown in Figures 3-10, 3-11and 3-12 is
   MTB> plot c4*c1;   c4 has the second coordinates, c1 has the first coordinates.
   SUBC> symbol;
   SUBC> type "x".

Topic 3.18: The Correlation Coefficient

The correlation coefficient, and its estimate, the sample correlation coefficient are only meaningful in the situation where both coordinates of the pair (x,y) are random quantities; not in the situation where x is under the control of the experimenter.

To get some understanding of these quantities, suppose we have the data pairs $(X_1, Y_1), (X_2, Y_2),...,(X_n, Y_n)$, where (X_j, Y_j) are a pair of values associated with the j-th individual, and not under the experimenter's control.

Definition 3.19: Sample Correlation Coefficient

Given a sample, $(X_1, Y_1), (X_2, Y_2),...,(X_n, Y_n)$, of data pairs, its **sample correlation coefficient**, $r_{x,y}$, is given by

$$r_{x,y} = \frac{1}{n-1} \frac{(X_1-\bar{X})(Y_1-\bar{Y}) + (X_2-\bar{X})(Y_2-\bar{Y}) +\cdots+ (X_n-\bar{X})(Y_n-\bar{Y})}{s_x s_y},$$

where $\bar{X}$ is the sample mean of the X_i, $\bar{Y}$ is the sample mean of the Y_i, s_x is the sample standard deviation of the X_i and s_y is the sample standard deviation of the Y_i (see Definition 3.13 on page 67).

❑

The population correlation coefficient, denoted by $\rho_{x,y}$, will turn out to be the value that the sample correlation coefficient gets close to as the sample size, n, becomes arbitrarily large. It can be shown that $-1 \le \rho_{x,y} \le 1$. The meaning of the correlation coefficient is probably not self evident when you first look at it. But notice that if the sample correlation coefficient $r_{x,y}$ is positive, then it seems likely that most of the terms $(X_i-\bar{X})(Y_j-\bar{Y})$ would be positive — which means that the two factors $(X_i-\bar{X})$ and $(Y_i-\bar{Y})$ would have the same algebraic sign; so that X_i would tend to exceed $\bar{X}$ precisely when Y_i exceeded $\bar{Y}$. The more positive $r_{x,y}$ is, the more likely it is for this tendency to occur. Conversely if $r_{x,y}$ is negative, X_i would tend to exceed $\bar{X}$ precisely when Y_i was less than $\bar{Y}$.

Based on this reasoning, it would appear that the correlation coefficient is a measure of how much X and Y *depend on each other* in any (X,Y) pair — but this interpretation is somewhat vague. A more *operational* interpretation is the following. Under very commonly made assumptions associated with normal distributions,[1] we can predict the value to be taken on by Y either **with** or **without** a knowledge of the value of X (without X the predicted value would be the population mean of Y).

The error in predicting Y using the value of X will, on the average, then be

$$\sqrt{1 - \rho_{x,y}^2}\ \sigma_y,$$ 3.6

---

1. These assumptions and the derivations which follow from them being covered in courses in Mathematical Statistics.

where σ_y (the population standard deviation of the Y_i) is the parameter representing the error of predicting Y without using knowledge of the value of X and $\rho_{x,y}$ is the population correlation coefficient.

So, if $\rho_{x,y} = 0.9$, we find $\sqrt{1 - \rho_{x,y}^2} = \sqrt{1 - 0.81} = \sqrt{0.19} = 0.436$.
Thus, if X and Y are 90% correlated, the error in predicting Y using X is only about 40% of what it would be if no use was made of X.

The formula for using the value of X to predict the value of Y is

$$\mu_y + \rho_{x,y}\, \sigma_y \frac{(X - \mu_x)}{\sigma_x}. \qquad\qquad 3.7$$

From this expression, it is seen that a change in X

of $c\,\sigma_x$ (i.e., of c σ_x- units)

leads to a change in the predicted value of Y

of $c\,\rho_{x,y}\,\sigma_y$ (i.e., $c\rho_{x,y}$ σ_y- units).

So the population correlation coefficient, $\rho_{x,y}$, is the **slope** of the line relating the predicted value of Y to the observed value of X — provided that the changes in X and Y are measured in their respective standard deviation units.

Example 3.11: Correlation Coefficient Usage

Suppose X is the natural logarithm of the measured systolic blood pressure of a randomly chosen 50 year old male (recall the discussion on page 74) and Y is the natural logarithm of the probability that this individual will have a heart attack sometime in the five year period following.[1] For now assume that the usual normality assumptions justifying the recently given formulae and interpretations are valid[2] for (X,Y).

In the situation being considered, the median blood pressure is assumed to be about 130 mm. of mercury, and thus the median of $X = log(SBP)$ (as well as the mean, due to the assumed normality of log(SBP)) is $log(130) = 4.8675$. We will assume that an increase of 60 mm of mercury over the median blood pressure brings

---

1. The reason for dealing with the logarithm of the blood pressure, rather than the blood pressure itself, is to get a better fitting model — a normal approximation to the blood pressure distribution is a poor one, since systolic blood pressure cannot be negative, and is almost never below 60 mm of mercury; on the other hand the systolic blood pressure can well be above 200 mm of mercury, but the median is probably near 130 mm of mercury; similarly probabilities must be between 0 and 1. For these reasons, it does not seem possible to get very good normal distribution fits to SBP or P(MI).
2. This is a purely *made-up* example, for illustrative purposes. The process of coming up with a usable model being illustrated is, however, similar to, but much more complex than what we do here.

you to the upper .995 point on the blood pressure distribution — i.e., the probability of a systolic blood pressure exceeding 190 mm of mercury is .005. (If you wanted truly trustworthy values here, you would have to examine some actual data.) Since the probability that a standard normal variable will exceed 2.33 is about .005, we see that this leads to the conclusion that

$$log(190) - log(130) = 2.33\sigma_x$$

from which we conclude that $X = log(SBP)$ has a normal distribution with population mean $m_x = 4.8675$ and standard deviation $\sigma_x = 0.16287$. It's worth drawing an accurate histogram for $SBP = exp(X)$. The histogram we see in Figure 3-13 does not appear too far out of line with reality, in the sense that it only presents values which are physiologically possible.

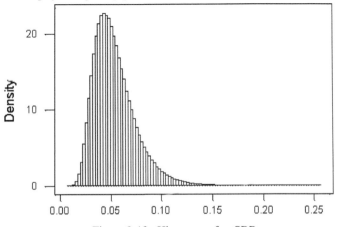

Figure 3-13   Histogram for SBP

The code for generating this histogram was the following.

```
MTB > random 500000 c1;
SUBC> normal 4.8675 .16287.
MTB > let c2 = exp(c1)
MTB > histogram c2;
SUBC> density;
SUBC> ninterval 100.
```

Similar reasoning, starting with the assumptions that the median probability of an apparently healthy 50 year old male having an MI within the next 5 years is .05, and that the probability that such an individual will have a probability of an MI within the next 5 years exceeding .12 is .005, we conclude that

$$Y = log(P(MI...)) \text{ has a normal distribution with}$$

$$\text{mean } m_y = -2.9957 \text{ and standard deviation } \sigma_y = 0.3757.$$

A density histogram of a sample of 500,000 from the distribution of P(MI...) is shown in Figure 3-14.

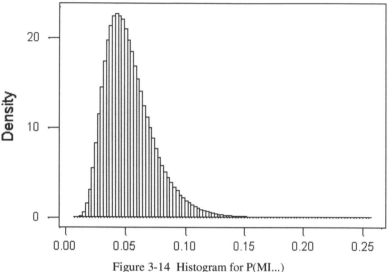

Figure 3-14 Histogram for P(MI...)

Now let us see what the predicted value of P(MI...) is if the measured *SBP* is 165 mm. of mercury and the population correlation coefficient, $\rho_{x, y}$, between $X = log(SBP)$ and $Y = log(P\{MI...\})$ is .4. In this case, the measured value of X is $log(165) = 5.1059$, and hence substituting $\mu_x = 4.8675$, $\sigma_x = 0.16287$, $\rho_{x, y} = 0.4$, $\mu_y = -2.9957$, $\sigma_y - 0.3757$, and $X = 5.1059$ into equation eqnum, we find that the predicted value of $Y = log(P\{MI...\})$ is -2.7757. But since exp(-2.7757) = .06231 [i.e., log(.06231) = -2.7757], we see that the predicted value of $P\{MI...\}$ is .06231. So with a correlation of .4 between $X = log(SBP)$ and $Y = log(P\{MI...\})$, we find that an increase of 65 mm of mercury above the median blood pressure leads to a predicted increase of .06231 - .05 = .01231 in the probability of an MI within the next 5 years. Moreover, the standard deviation of the predicted $Y = log(P\{MI...\})$ is now $\sigma_y\sqrt{1 - \rho_{x, y}^2} = \sigma_y 0.917$, so that the deviation of the predicted value from the unknown true value tends to be only about 90% of the corresponding deviation of the prediction made when X is not taken account of. This translates into an improved prediction for $P\{MI...\}$.

To summarize — under the given model, with a correlation coefficient of .4 between the normally distributed quantities $log(SBP)$ and $log(P\{MI...\})$, we predict $P\{MI...\} = 0.06231$ when $SBP = 165$. This prediction tends to be more accurate than the prediction of $P\{MI...\}$ when SBP is not taken into account. ∎

It should be realized that it is not legitimate to conclude from this type of result, that reducing the blood pressure by some artificial means (such as leeches or drugs) will reduce the probability of a heart attack. Some such reductions might work, while others might produce exactly the opposite effect. It could well be that there is some underlying cause which produces both an elevation in blood pressure and a tendency toward heart attacks. Just lowering the blood pressure might not affect the root cause or may even aggravate it. The best way to investigate whether a change in X causes a change in Y is by means of a controlled experiment, where we do our best to ensure that the only difference between the two groups in the test is a controlled change in X.

Practice Exercises 3.12

1. Based on two separately generated standard normal data sets, each of size 10,000 the first set put in column c1, and representing repeated observations on some quantity X and the second set (put in column c2) and representing repeated observations on some other quantity Y.

   a. determine the sample correlation coefficient $r_{x,y}$ using the command   *correlation c1 c2*.
   What does this result seem to indicate about how useful X is in improving our prediction of Y?

   b. Let $Z = X + Y$.   How would you use the value of X to improve the prediction of Z? *Hint*: Let c3 = c1 + c2.

   c. Let $W = 11X + 2Y$. Discuss the prediction of W using nothing at all; using only X; and using only Y.

2. Determine what happens to the predicted values of $P\{MI...\}$ as $\rho_{x,y}$ and *SBP* take on various other values.

3. Let X have a standard normal distribution and let $Y = X^2$.

   a. Is Y *dependent* on X? (Does knowing the value of X improve the predicted value of Y?)

   b. Generate 5 samples of size 25 of X and of Y, putting them in columns c1 - c5 and c6 - c10, respectively. Compute the 5 sample correlation coefficients, via correlation c1 c6, correlation c2 c7,..., correlation c5 c10, respectively. Explain the results you observe.

Topic 3.20: The empirical cumulative distribution function (EDF)

The empirical cumulative distribution function (EDF), unlike the histogram, is a descriptive statistic which is not usually used by itself to discover attributes of the underlying population. But it can be a valuable tool to check the validity of common assumptions, such as that of normality of data. The single variable EDF, usually denoted by the symbol F_n is a function determined by a numerical data sequence $(x_1, x_2,..., x_n)$. Its value, $F_n(x)$ at any number x is given by the formula

$$F_n(x) = \text{proportion of the elements of} (x_1, x_2,..., x_n) \leq x. \qquad 3.8$$

- If the sequence, $(x_1, x_2,..., x_n)$, represents repeated independent observations of some random quantity X (such as repeated LDL measurements) then $F_n(x)$ is an estimate of the probability that $X \le x$. This probability is denoted by $F(x)$, and the function F is called the *(cumulative) distribution function* (CDF) *of X*.

If we hypothesize that the underlying distribution of the data being investigated has a standard normal distribution, we might overlay that graph of the standard normal CDF and the EDF. We illustrate this for two cases, the first where the data has a standard normal distribution, and the second where it has a uniform distribution on 0 to 1. Later we will introduce some formal tests using the EDF.

What follows is the macro, *empdis.mac*, used to generate the overlaid graphs of the EDF and hypothesized CDF. A few comments on this macro may help in understanding the code. The original data points, from which the graph is to be plotted may look as shown in Figure 3-15 .

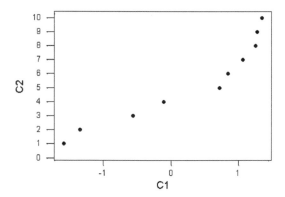

Figure 3-15 Original Data Points

The code which generated this graph is the following.

```
MTB > random 10 c2
MTB > sort c2 c1
MTB > set c2
DATA> 1:10/1
DATA> end
MTB > plot c2*c1
```

Unfortunately the graph we want is somewhat different from what you see in Figure 3-15. By adding additional points to the data generating this graph, shown by the additional xs in Figure 3-16, arranging the points in order from left to right, with lower coming before upper, and using the *connect* subcommand we can get what we want. This is done within the macro *empdis*.

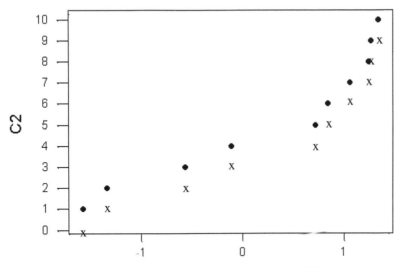

Figure 3-16 Added Points to Construct EDF

Before running the macro *empdis*, we put the data in column c2, and points from the hypothesized distribution in columns c8 (first coordinates) and c9 (second coordinates). It is important that the CDF being overlaid with the EDF have the same minimum and maximum first coordinate values (in order that the graphs be printed with the same scale in Minitab; some other programs take care of this problem automatically). This is done as follows. After loading the data sequence $(x_1,...,x_{k1})$ into column c2, we type

let k10 = min(c2) - 1.2

let k11 = max(c2) + 1.2   (The value 1.2 is the extension of the EDF graph.)

Then, if it looks like .1 is a reasonable first coordinate spacing, we type the commands

```
set c8
k10:k11/.1
end.
```

This puts the proper first coordinate values in column c8. If we want to put a standard normal CDF in column c9, we type

```
cdf c8 c9;
normal 0 1.
```

What we get when we run the macro *empdis* (via typing the command %c:\stbk\macs\empdis, if this macro was stored in the directory c:\stbk\macs) is shown in   Figure 3-17.

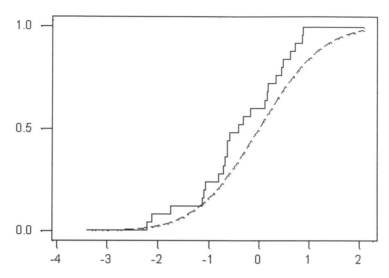

Figure 3-17   Overlay of Standard Normal CDF and EDF
                          Based on 25 Standard Normal Data Points

Here is the exact code (together with the Minitab responses) we typed for the first such EDF CDF overlay, where the data was from a standard normal distribution, and the CDF was a standard normal CDF, resulting in Figure 3-17 above.

```
MTB > let k1 = 25
MTB > random k1 c2
MTB > let k10 = min(c2) - 1.2
MTB > let k11 = max(c2) + 1.2
MTB > set c8
DATA> k10:k11/.1
DATA> end
MTB > cdf c8 c9;
SUBC> normal 0 1.
MTB > %c:\stbk\macs\empdis
Executing from file: c:\stbk\macs\empdis.MAC
```

If instead we had put in a different type of data, as in the following code,

```
MTB > let k1 = 25
MTB > random k1 c2;
SUBC> normal 3 2.
MTB > let k10 = min(c2) - 1.2
MTB > let k11 = max(c2) + 1.2
MTB > set c8
DATA> k10:k11/.1
DATA> end
MTB > cdf c8 c9;
```

```
SUBC> normal 0 1.
MTB > %c:\stbk\macs\empdis
Executing from file: c:\stbk\macs\empdis.MAC
```

here is the graph that would have resulted.

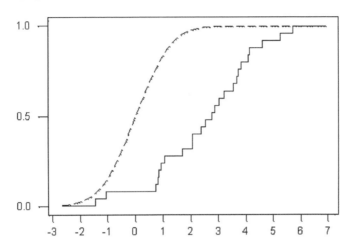

The code for the macro *empdis* is given in the segment which follows. It is documented so as to make it understandable and to allow easy modification.

GMACRO

empdis

the original unsorted data x1,...,xk1 is put into column c2

let k2 = k1 -1

sort c2 c1   # the sorted original data is put into column c1

set c2

0:k2/1

end

do k2=1:k1 #this loop is to get the extra points needed in the plot

let k4=2*k2 - 1

let k5 = k4 + 1

let c3(k4) = c1(k2)

let c3(k5) = c1(k2)

let c4(k4) = c2(k2)   # These two commands create the two second
                          # coordinates above each first coordinate for
                          # plotting successive points.

```
let c4(k5) = c2(k2) + 1

enddo


let k6 = k5 + 1  # These 3 commands extend the EDF graph 1.2 to the right,
                 # for aesthetic reasons only.
let c4(k6) = c4(k5)
let c3(k6) = c3(k5) + 1.2


let k7 = 2*k1 + 1  #k7 is the number of points in c3 (as well as in c4).
    #This next loop is to shift the points, to allow the EDF graph to extend left.
do k8 = k7:1
let k9 = k8 + 1
let c4(k9) = c4(k8)
let c3(k9) = c3(k8)
enddo # Now each element of both c4 and c3 have been copied 1 row down.
    # The first elements of c4 and c3 are currently the same as the 2nd ones.
let c3(1) = c3(2) - 1.2 #This extends the EDF 1.2 units to the left — again
                        #for aesthetic reasons.
 let c7 = c4/k1        #This normalizes the EDF (max value = 1).
plot c7*c3 c9*c8;      #c9,c8 have the theoretical CDF.
overlay;
connect.
ENDMACRO
```

Practice Exercises 3.13

Run the *empdis* macro for the following data sets of sizes 5,10,25 and 60 values respectively and hypothesized CDF's

	Data	Hypothesized CDF
1.	Normal 0, 1	Normal 0, 5
2.	Normal 0, 1	Normal 0, .2
3.	Normal .5, $1/\sqrt{12}$	Uniform 0, 1
4.	Normal 1, 1	Poisson 1
5.	Normal 1, 1	Poisson 3

For each of the above, in which of the cases does the data not appear to come from the hypothesized distribution?

Problems 3.21

1. Using (and if necessary, appropriately modifying) the *empdis* macro just presented, get comfortable with how the EDF relates to the CDF which gave rise to it, and to other CDFs. Alternatively, vary the data, from arising from the given CDF, and arising from other CDFs. These exercises should prepare you for the more formal results represented by the Kolmogorov-Smirnov tests of fit.[1]

2. Modify the *empdis* macro to overlay two EDFs. Such graphs should prepare you for the Kolmogorov-Smirnov two-sample tests.[1]

Review Exercises 3.14

1. What are the main reasons for sampling?

2. Why is sampling, as opposed to a complete examination of a lot of manufactured items:

   a. often desirable

   b. sometimes unavoidable in determining whether the lot is a satisfactory one?

3. Explain the importance of random sampling.

4. How do you implement random sampling to obtain test objects for determining which of two manufacturing methods is superior?

5. What is the purpose of random sampling in the previous exercise?

6. Provide some situations in which

   a. the mean seems more appropriate (useful) than the median.

   b. the median seems more appropriate than the mean.

7. Does a larger sample size almost always yield more accurate results in estimating a population mean than a smaller one? Explain your answer.

8. What are the advantages of a larger sample size in estimating some quantity, such as the population mean?

9. Use the macro *meanseq* to get an idea of how many observations are needed until the sample standard deviation of normal random variables with a population standard deviation of 3 are likely to be within .2 units from this population value.

10. What is the two-sigma rule-of-thumb?

---

1. These tests, not currently available on Minitab, are available on some other packages. A macro with these tests that can be downloaded from the CRC website will be made available.

11. Estimate how accurate the two-sigma rule-of-thumb is for the following distributions

    a. The normal distribution with mean 2 and standard deviation 5.

    b. The Poisson distribution with mean .2

    c. The Poisson distribution with mean 4

    d. The Poisson distribution with mean 40.

    e. The binomial distribution with $n = 20$ and $p = $ .01, .1, .3, and .5

    f. The uniform distribution on -.5 to .5

    g. The distribution of $X + Y$ where these variables are uniform on −.5 to .5, and statistically independent (The independence here has the effect of letting you generate independent repetitions of this sum by using the top menu bar Calc -> Random Data feature of Minitab.)

    h. The distribution of $X + Y + Z + U$ where these are independent and uniform on −.5 to .5

12. Suppose the probability of a child having leukemia at some specified point in time is .00004, and suppose the distribution of the number of children with leukemia in a group of 1,000,000 children is Poisson. This would mean that the mean number of children with leukemia in this group is 40. Suppose that, in some city of 1,000,000, the observed number of children who had leukemia was 75. Does this result lead you to believe that there is a problem in this city?

    Hint: Compute the probability of observing a value of 75 or more from a Poisson distribution with mean 40.

    a Same as previous question, but using the results of Exercise 11.d above.

Chapter 4
Survey of Basic Probability

4.1 Introduction

The purpose of this section is to present convincing reasons for why a solid theoretical probability foundation is needed in the development of statistics. We already know that the subject of statistics is concerned with the development and application of statistical models — descriptions in which there are significant factors affecting what is observed, but which are not accounted for explicitly. In the previous chapters we developed a qualitative approach to getting familiar with the behavior of such models by looking at the behavior of data sets drawn from populations with known characteristics. This approach is vital in learning how their properties are reflected in samples drawn from these populations. However, this approach alone is not adequate to properly answer many of the questions that come up, because it can only examine a finite number of cases.

At the very least, you need to know whether you have examined sufficiently many cases to effectively cover all those of interest — and for this you need some an adequate justifying theory. Most scientists plot a finite number of points to represent the graph of the functions they are investigating. To be sure that of the legitimacy of linear interpolation between plotted points, trustworthy theory is required.

Using simulation you can certainly get a decent idea of how many experiments must be performed in order to estimate the probability of a successful one to within .02, 95% of the time. But this information doesn't make it at all obvious how many experiments to perform to be 99% sure that your estimate of the probability of success is within .001 of the *true* value.

Finally, and maybe most important, the empirical approach by itself does not furnish us with good yardsticks to determine the adequacy of the Monte Carlo results we obtain. To give us a reasonable amount of faith in simulation, at the very least we must demonstrate that this method gives satisfactory results for similar problems where we know the theoretical answer. And for this, we need to know how to find theoretical answers to a wide range of problems.

A solid foundation of probability theory is needed

- to determine which parameters in the model correspond to observed sample quantities, and to determine something about how far the sample quantities are likely to be from their model counterparts

- to provide theoretical confirmation of behavior that has been observed in Monte Carlo simulations

- to design statistical procedures which have the characteristics we want — one special case being obtaining estimates which have a prescribed accuracy

- to obtain dependable *yardsticks* for judging the adequacy of computer simulations, and for designing these simulations

- to determine the behavior of statistical test procedures which have been devised

- to extract information about a system from a chosen statistical model.

Here are some illustrations to help clarify what has just been claimed.

Example 4.1: Population means

Given the formula for some probability function or probability density, e.g., a Poisson probability function (see page 31), what is a formula for the population mean and for the population standard deviation — the values approached by the sample mean and sample standard deviation respectively, for large sample sizes? It can be shown theoretically that for a Poisson distribution the population standard deviation is the square root of the population mean. From this, it can be shown theoretically that the relative error in estimating the population mean tends to be small for large Poisson population means. This result might be reasonably conjectured from Monte Carlo simulations. The theory not only establishes it, but provides quantitative results as to how large this relative error tends to be. Such results are very useful when we want to know the error in our estimates of quantities such as the average number of phone calls initiated in a given time period, or the probability of more than a specified number of aircraft arrivals in some time interval.

The Central Limit Theorem (Theorem 4.38 on page 135) shows that under very general conditions sample means tend to have CDFs *close*[1] to a normal distribution CDF. This result is especially useful for quick simple approximations to probabilities involving sample means.

∎

Example 4.2: Sample size needed to estimate probabilities

Based on Monte Carlo simulations, we can get a good idea of how many observations are needed to accurately estimate the probability of success of an experiment with a specified probability. It is reassuring that we can theoretically derive a formula for the number of observations needed to estimate any specified probability to a given accuracy with high assurance of drawing a correct conclusion. Even more significant, we can determine the number of observations which guarantee that no matter what the probability, our estimate has high assurance of some guaranteed accuracy. In fact, it is possible to find a number such that with

---

1. In a sense to be specified. In some cases, particularly independent identically distributed uniform variables, the density of their normalized mean is remarkably close to the standard normal density, even for very small sample sizes.

specified assurance, we can estimate the entire CDF of any random quantity to a given accuracy. Without the theory our conclusions would have much less credibility.

■

Example 4.3: Estimating standard deviations

Some problems are tougher to handle than the ones just presented. Theory shows that generally we cannot decide in advance on a fixed number of observations which will guarantee that we can estimate standard deviations to within some specified accuracy, even if we know that the data is normally distributed. In this situation, the theory shows that you **can** get the information you want, but it has to be done in two stages — a pilot stage to get a ballpark estimate of the standard deviation, which then lets you decide on the number of observations needed in the next stage.

■

Example 4.4: Distribution of the maximum

In dealing with flood control we tend to look at quantities such as the maximum flood over a 50 year period. Suppose the number of floods in this period is denoted by n, and $X_1,...,X_n$ were the corresponding flood crests. A critical element in getting a handle on the probability distribution of the worst such flood consists of noting that the worst such flood has a crest not exceeding f feet if and only if

$$X_1 \leq f \text{ and } X_2 \leq f \text{ and } \cdots \text{ and } X_n \leq f,$$

whose probability can then often be determined using standard techniques in this field.

■

Example 4.5: Behavior of sample percentiles

We know that percentiles such as the population median and quartiles provide useful information about the underlying population (see page 60). It is reasonable to conclude from Monte Carlo simulations that for large samples, the sample percentiles and sample quartiles get close to these population parameters if the population is described by nonzero probability density functions. It isn't hard to find the CDF of any sample percentile, e.g., the sample median, in this case; for simplicity assume that there are an odd number, $2n + 1$ of observations, all distinct. Then the probability that

$$sample \ median \leq x$$

is just the probability that at least n of the observations are less than or equal to x. This is given by 1 - binomial CDF at $n - 1$, based on $2n + 1$ observations with $p = F(x)$, where F is the CDF of each of the observations.

So once again the theory of probability is needed to determine the distributions of the sample percentiles. Not only is this useful as a yardstick for judging the adequacy of our Monte Carlo methods, but it will turn out to enable us to construct confidence intervals for these percentiles.

■

Example 4.6: Characteristics of statistical hypothesis tests

Tests of hypotheses occupy a good deal of space in medical journals. For instance tests of whether or not some pollutant is a carcinogen, or tests of whether or not a new treatment is better than the accepted standard, To decide whether or not a given test procedure is worth using, when we have specified the amount of data that can be examined, we look at the probability of rejecting a specified one of the two given hypotheses (this being called the *null hypothesis*) under the various different possible scenarios — for instance we would like to know the probability of rejecting the null hypothesis for the following possible situations:

Old standard's cure probability	New treatment cure probability
.1	.1
.1	.2
.2	.2
.2	.3

etc., in order to see how good a *discriminating power* the proposed test has.

In fact, **the worth of just about every known statistical procedure is judged in terms of probabilities which must be calculated** (showing just how much the field of statistics stands on a foundation of probability theory).   ∎

4.2 Probability and its Basic Rules

The aim of probability theory is to enable us to compute probabilities of events of interest and related quantities, such as:

- the probability of no failures in 20 runs of an experiment with some given probability of success

- the probability of rain today

- the mean time between failures of your computer

- the probability of developing a deadly mutation from exposure to the sun, or to various pollutants

After deciding on the basic rules of the subject, probability may be thought of as consisting of two parts:

mathematical probability and **applied probability**.

Mathematical probability is concerned with using the rules of probability together with given assumptions about certain probabilities and associated quantities, to determine some other related probabilities or quantities. The classical example of this is to determine the probability of 60 successes in 100 independent trials of an experiment whose probability of success is .25.

Another example of mathematical probability is establishing the relationship between the binomial and normal distributions (which yields a simple approximation to the solution of the preceding computation).

Applied probability is devoted to all of the other aspects of investigating particular real-life probability problems, problems such as:

- determining how to take account of partial information about the results of an experiment — for instance, how to revise your election predictions in light of partial returns

- determining how to define the population mean value in terms of the discrete probability function in such a way that it corresponds to what the sample mean gets close to for large sample sizes

- determining how to characterize the concept of *independence* (and other concepts as well) in terms of the characteristics that such probability distribution must exhibit

- how to make appropriate choices of model characteristics to correspond to properties which should be exhibited, e.g., a tendency for one measured variable on an object to increase quadratically as another such variable increases.

In a sense, applied probability is concerned with customizing the model you choose so that it does what it should, namely, reflect reality.[1]

We will shortly begin the serious business of developing the basic rules of probability. They turn out to be pretty simple, which may leave you unprepared for just how complicated things can sometimes get. What's important is to avoid getting bogged down in some of the messier details, and to keep focussed on the simple ideas which give rise to them.

We note that most of us use the terminology of probability in everyday life. We talk about the event that the weather is sunny, about the occurrence of this event and about the probability of (the occurrence of) this event. Although we might be able to develop the theory of probability with no additional framework, it turns out to be a little bit easier to have a more concrete setting for the theory. The setting we use was introduced by the twentieth century Russian mathematician, Kolmogorov, and is centered around a set-theoretic model for an experiment.

When we think about an experiment, we should be aware that experiments usually require measuring instruments to record their results. These instruments may consist of our own senses, and usually other devices such as voltmeters (to measure voltage [*electrical pressure*]), clocks (for the necessary time measurements), rulers or other fancier distance measuring devices, etc. The measuring instruments used in an

---

1. As such, statistical inference — narrowing down the class of possible models with the help of data (see Chapter 2 Section 2.3, Chapters 5, 6, and 7) — may be thought of as part of applied probability.

experiment limit the possible results that can be observed from the experiment. The instruments associated with an experiment (at least conceptually) specify the so-called *sample space*, which is usually thought of as *the set of all possible outcomes* of the experiment.[1]

Definition 4.1 Sample space

The *Sample Space* for an experiment, frequently denoted by the symbol S, is a set which includes all of the possible outcomes for this experiment.

❏

The elements of the sample space for an experiment can be thought of as dictated by the measuring devices being used for recording experimental results.

Example 4.7: Sample space for toss of a pair of dice

One possible sample space consists of the numbers 2,3,4,...,12. As it turns out, for the purposes of assigning probabilities, a better sample space is the set of ordered pairs given by Table 1.1 on page 8.

■

Next we must define what we mean by an event.

Definition 4.2: Event, occurrence of a given event

An *event* is a subset of the sample space, to which a probability is assigned.[2]

If we denote some subset of the sample space by the symbol E, we say that the event E has occurred in a particular run of the experiment, if on this run the observed outcome was an element of the subset E.

❏

Example 4.8: Events in tossing a pair of dice

Using Table 1.1 on page 8 as the sample space, S, for the toss of a pair of dice,

- the event *Sum of* 3 is the subset $\{(1,2),(2,1)\}$ of S.
- the event *FD2: First die comes up 2* is the subset $\{(2,1),(2,2),(2,3),(2,4),(2,5),(2,6)\}$

If the observed outcome was (2,4) then the event *FD2* occurred.

If the observed outcome was (5,2) then *FD2* did not occur.

---

1. Actually, to avoid inconvenience in the mathematical development, the sample space frequently includes objects which could not under any circumstances, represent possible experimental outcomes. This will occur naturally and often, and will not usually cause any difficulties.
2. Because of the rules for assigning probabilities, it turns out that there are situations in which not all subsets of the sample space qualify as events.

It is **a common error** to claim that occurrence of *FD2* means that all outcomes in *FD2* occurred (whatever this would mean). **If *FD2* has occurred, then the outcome is one of the elements of the subset**

$$\{(2,1),(2,2),(2,3),(2,4),(2,5),(2,6)\}.$$

Notice that if we had chosen the set $\{2,3,4,5,6,7,8,9,10,11,12\}$ as the sample space for tossing a pair of dice (each element representing a possible sum), then *FD2* **could not be considered an event in this sample space**.

∎

The above example shows that **we must be careful when choosing the sample space, *S*, to ensure that all events of interest correspond to subsets of *S*.** This has to be a conscious choice on our parts, because it doesn't happen automatically. (Put another way, we need to choose our measuring instruments carefully in order to get the information we want.)

Topic 4.3: Formation of events from other events

The next step in our development is to see how to compute the probability of an event formed from other events whose probabilities are known. But first we must ask, *How can events be formed from other events?* It is here that we see the advantages of letting events be subsets of the set called the sample space; because there are natural ways to form sets from other sets. We first do this informally.

One way to create a different subset from a given one is the following. Given any event A (subset of the sample space S) another subset (if S has at least one member) is the set of all members of S which are **not** in the subset A. This subset is called the *complement* of A (with respect to S — this modifier usually being omitted, even though it is understood) and denoted by A^c. As will be seen, it is sometimes much easier to compute the probability of the complement of some event, than it is to compute the event's probability directly. Then, by the frequency interpretation of probability (page 19), since

proportion of observations in A + proportion of observations outside $A = 1$,

we would expect

$$P(A) = 1 - P(A^c),$$

where $P(A)$ denotes the probability of A, and $P(A^c)$ the probability of A^c.

Another way of forming events (subsets) from other events is taking their union. Here the union of any collection of sets is just the collection whose members are members of at least one set in the original collection. If the collection of sets just consists of two sets, A and B, their union is denoted by the symbol $A \cup B$. In general, if for each value t in some (index) set T, A_t stands for a set (of possible outcomes) then the union $\bigcup_{t \text{ in } T} A_t$ is the set of outcomes which are members of at least one of the A_t.

Everyone knows what is meant by the intersection of two streets - it consists of the locations on both of the streets. The intersection, $A \cap B$ is the set of outcomes in both of the events A, B. The intersection $\underset{t \text{ in } T}{\cap} A_t$ of all of the specified A_t, is the event consisting of those possible outcomes which are members of all of the A_t.

Example 4.9: Dice illustrations of union, intersection

Using the 36 point sample space for tossing two dice (Table 1.1 on page 8) let

    7 represent the event   {(6,1),(5,2),(4,3),(3,4),(25),(1,6)}
    11 represent the event   {(6,5),(5,6)}
    The union, *7* $\cup$ *11*, of these events (usually called *7 or 11*) is the event
        {(6,1),(5,2),(4,3),(3,4),(2,5),(1,6),(6,5),(5,6)}.

Let *>9* represent the event {(6,4),(5,5),(4,6),(6,5),(5,6),(6,6)}.

The union *>9* $\cup$ *7* $\cup$ *11* is the event
{(6,4),(5,5),(4,6),(6,5),(5,6),(6,6),(6,1),(5,2),(4,3),(3,4),(2,5),(1,6),<u>(6,5),(5,6)</u>}

Note that the underlined elements are duplicates and can be deleted (because a set is determined by its members).
                                                                      ■

Before we go on to provide some more interesting examples of unions, complements and intersections, it seems worthwhile to illustrate the concepts with so-called *Venn diagrams*. These illustrations are very helpful in suggesting both some of the basic rules governing the assignment of probabilities, but also help to motivate some of the later results of probability as well as certain set manipulations (DeMorgan's Laws) which we will be using to get simple, useful bounds for probabilities of unions and intersections.

Here we represent the sample space S as a large ellipse (with its interior).

The points in S represent distinct possible outcomes. Events in the sample space are represented as smaller figures, as shown in Figure 4-1 .

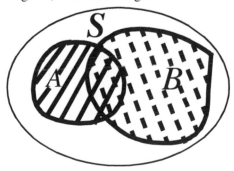

Figure 4-1 Venn Diagram of Sets A, B

A is represented by the shading *///*, B is represented by ❜❜❜ .

$A \cup B$ is shown in , while $A \cap B$ is the cross-hatched part of $A \cup B$.

The complement of the event A is indicated by the shaded area in Figure 4-2, with the sample space, S, being the entire ellipse.

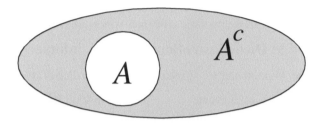

Figure 4-2 Set A and its Complement

Now we summarize the concepts just introduced.

Definitions 4.4: Formation of events from other events

Suppose A,B and for each t in the *index set* T, A_t are events in the sample space, S of some experiment (see Definition 4.1 and Definition 4.2 on page 94).

The **union** of the two events, A and B, denoted by the symbol $A \cup B$, is the set of possible outcomes in at least one of these events. In general the union, $\bigcup_{t \text{ in } T} A_t$ is the set of outcomes which are members of at least one of the A_t.

The **intersection**, $A \cap B$, is the set of outcomes in both of the events   The complement, A^c, of the event A (relative to S) is the set of members of S which are not in A. Note that $(A^c)^c = A$.

The set S^c is the *empty set* — the set having no members. At first sight this might not seem worth mentioning. However, when an event cannot possibly ever occur, it must correspond to the empty set. Having a symbol, ϕ, for the empty set (just as having the symbol 0 for the number of members of the empty set) turns out to be quite convenient in probability — for instance when $A \cap B = \phi$ this indicates that A and B have no possible outcomes in common, that is, they are mutually exclusive events.

❏

Example 4.10: Convolutions

The simplest convolution, which arises in dice tosses, is probably the best one for understanding this concept. It arises as follows. Letting the sample space consist of the 36 points in Table 1.1 on page 8, let

F_i be the event "roll of first die results in the value i"

S_j be the event "roll of second die results in the value j."

The event that the sum on the two dice is k may be written as the *convolution*

$$\bigcup_{\substack{\text{all } i \text{ from} \\ 1 \text{ to } 6}} F_i \cap S_{k-i}$$

because a sum of k results if and only if for some integer value i from 1 to 6, the first die comes up with the value i and the second one comes up with the value k-i.

Now, suppose we have an integer valued experiment, such as determining the number of failures of a computer during some specified period of operation. Looking at two computers, if

F_i is the event that the first computer fails i times during this period,

S_j is the event that the second computer fails j times during this period then

the event that the total number of failures of the two computers is k may be written as the convolution

$$\bigcup_{\substack{\text{all integer } x \\ \text{with } 1 \le x \le 6}} L_x \cap P_{k-x}$$

Now suppose the possible outcomes of our experiment consist of ordered pairs or real numbers, (X, Y), where X represents the error on last year's estimate of GNP (gross national product) and Y the error on GNP of a year earlier. Then X - Y is how much the error has changed. Let L_x be the event that the error last year was the given number x, and P_y the event that the previous year's error was the given number y. The event that the difference of these errors had the value z may be written as the convolution

$$\bigcup_{\text{all real } x} L_x \cap P_{z+x}$$

The event that X - Y is less than z may be written

$$\bigcup_{\text{all real } q < z} \left(\bigcup_{\text{all real } x} L_x \cap P_{q+x} \right)$$

■

It is probably not worth trying to get too proficient with the set manipulations just presented, because the ones you are likely to learn are the ones that you use frequently — and you won't know what they are until you get fairly deeply involved in some real problems. Nonetheless, here are some problems for those who want to get more feel for this area.

Problems 4.5

1. Convince yourself of the following equalities.

   a. For all events A,B, $A \cap B = B \cap A$

   b. For all events A,B, $A \cup B = B \cup A$

   c. For all events A,B,C, $A \cap (B \cup C) = (A \cup B) \cap (A \cup C)$
      Hints: Draw some pictures.

   d. Convince yourself of the following (called *De Morgan's Laws*).
      If for all t in the index set T, A_t and B_t are subsets of S, then

$$\left(\bigcup_{t \text{ in } T} A_t \right)^c = \bigcap_{t \text{ in } T} A_t^c \quad \text{and} \quad \left(\bigcap_{t \text{ in } T} B_t \right)^c = \bigcup_{t \text{ in } T} B_t^c$$

Hints: For the first equality, if you are outside the union, you must be outside each element of the union (since if you were in even one of the elements, you'd be in the union). Hence for each t you must be in the intersection of the complements of each union element.

For the second equality, let $A_t = B_t^c$ in the first equality and take complements of both sides. Some pictures might be helpful also.

2. Suppose S is the set of ordered pairs (x,y) of real numbers — the X-Y plane. For all real numbers a, b, let

L_a be the set of points (x,y) in S such that $x = a$

M_b be the set of points (x,y) in S such that $y = b$

a. Let d stand for a real number, and let Q_d stand for the set of points (x,y) in S such that $xy = d$.
Write Q_d in terms of the sets L_a and M_b.
Hint: You should consider two cases: $d = 0$, and $d \neq 0$

b. Let d stand for a real number, and let R_d stand for the set of points (x,y) in S such that $xy \leq d$.
Write R_d in terms of the sets L_a and M_b.
Hints: You might want to consider three cases, $d < 0$, $d = 0$ and $d > 0$.
Try to draw the sets R_d for various values of d, both from this representation, and directly from the definition. It's a bit tricky.

The rules of probability we are now going to present are motivated by the frequency interpretation — where we think of the probability of an event as the (long-run) proportion of occurrences of this event in many independent repeated trials. This immediately leads to the rule that all probabilities lie in the interval from 0 to 1. Both 0 and 1 are allowable probabilities. In fact, because of its definition, the proportion of occurrences of the sample space, S, is 1 (100%). Hence another postulated rule of probability is that $P(S) = 1$.

You might ask what happens if something else occurs, such as the coin landing on its edge, thus producing neither *Heads* nor *Tails*. Then we would not consider this a proper performance of the coin-toss experiment; which does not mean we would necessarily ignore it — but simply that the result does not fit the type of analysis we are carrying out. When the Michelson-Morley experiment to measure the speed of light through the (nonexistent) ether gave a totally unexpected result, far from being ignored, it led Einstein to the special theory of relativity

Since the proportion of occurrences of the union of two events which have no members in common is just the sum of the proportions of their occurrences, we postulate that if two events, A and B, have no possible outcomes in common, then

$P(A \cup B) = P(A) + P(B)$. For reasons of mathematical convenience, Kolmogorov actually made this rule stronger, by requiring that if $A_1, A_2,...,$ is any finite or infinite **sequence** of events, not two of which have any elements in common, then the probability of their union is the sum of their probabilities.

We summarize the basic rules we will be using as follows.

Definition 4.6: Probability Measure

A probability measure, P, is a function which assigns probabilities $P(A), P(B),...$ to events $A, B,...$. The sample space, S, the complement, A^c, of any event, A, and the union as well as the intersection of any finite or infinite sequence[1] of events, $A_1, A_2,...$, is an event.

$$P(S) = 1.$$

Furthermore if the sequence $A_1, A_2,...$ consists of mutually exclusive events $(A_i \cap A_j = \phi$ for $i \neq j)$, then[2]

$$P\left(\bigcup_{\text{all } i} A_i \right) = \sum_{\text{all } i} P(A_i)$$

Note that although the union of a sequence of events is an event, even knowing the probabilities of each of the events in this sequence does not usually permit computation of the probability of their union. More information is usually needed, such as that these events are mutually exclusive. In our illustrations it will become clear that the probability of the union of two events is only rarely determined by their individual probabilities alone.

❑

That's it — these are the only universal rules for assigning probabilities. So you might think their simplicity would make probability an easy study. Not so. You can

---

1. When we refer to a sequence of objects, we are assuming that for each counting number, i, in some set, C, of counting numbers, there is an associated object, say A_i. If the set, C, of counting numbers includes more elements than can be described by any given counting number (e.g., more than 10, more than 100, ... etc.) the sequence A_i for i in C is called an *infinite* sequence. The correct way to characterize such a sequence, C, is that for each counting number, n, its elements cannot be put into a one-to-one correspondence with the sequence $(1,2,...,n)$. So the set of even counting numbers is an infinite sequence.
    In the case of an infinite sequence — this sum is the sum of an infinite series, namely the value the finite partial sums approach with inclusion of many terms. For those not familiar with the theory of infinite series, it is usually adequate just to think in terms of sums of a finite number of terms.

2. The symbol $\sum_{\text{all } i} P(A_i)$ indicates adding up all of the terms $P(A_i)$ — the symbol $\sum$ being the upper case Greek letter sigma (Greek S, as in *sum*)

see the fallacy of this reasoning by noting how difficult it would be to build a house if your only tools were a hammer, nails, and a hand-saw.

Here is a fairly obvious result which follows from the rules for probability (Definition 4.6), namely for all pairs of events, A, B

$$P(A \cup B) = P(A) + P(B) - P(A \cap B).$$ 4.1

Equation 4.1 is *fairly obvious* because if you think of probability as a proportion, when you add $P(A) + P(B)$, each point in the overlap, $A \cap B$, gets accounted for twice, while the other elements of $A \cup B$ get accounted for only once. So, if you want to compute $P(A \cup B)$ after adding $P(A) + P(B)$ you must subtract $P(A \cap B)$.

Equation 4.1 can be proven mathematically, but you might wonder, if this result is so obvious, why should we even bother proving it? The answer becomes clear if you ask yourself, *suppose it could not be proved?* If that were the case, then you would be justified in feeling that the mathematical model set up in Definition 4.6 had no worthwhile purpose. (After all, if it can't even establish obvious results, how could it help with more challenging problems.) Our proof will show that Definition 4.6 does embody at least some of what we expect of any probability assignment. The steps in the proof are quite reasonable (and are most easily understood if you draw pictures to illustrate each set-manipulation step). Here they are.

We first write A as the union of the part of A in common with B and the part of A disjoint from B. That is

$$A = (A \cap B) \cup (A \cap B^c)$$ 4.2

as we see in Figure 4-3 below.

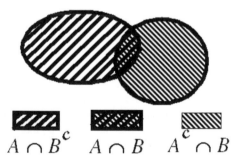

$$A \cap B^c \quad A \cap B \quad A^c \cap B$$

Figure 4-3 Decomposition of $A \cup B$ into a Union of 3 Disjoint Events

Notice from Figure 4-3, the events $A \cap B$ and $A \cap B^c$ are mutually exclusive, so that from the basic rules of probability from equation 4.2 we have

$$P(A) = P(A \cap B) + P(A \cap B^c),$$

or

$$P(A \cap B^c) = P(A) - P(A \cap B)$$ 4.3

Also notice from Equation 4.2 and some evident rules for set algebra we have

$$A \cup B = B \cup A$$
$$= B \cup ((A \cap B) \cup (A \cap B^c))$$
$$= B \cup (A \cap B^c)$$

Since this last union is a union of mutually exclusive events, from the rules given in Definition 4.6 we find that

$$P(A \cup B) = P(B \cup (A \cap B^c)) = P(B) + P(A \cap B^c)$$

which we rewrite as

$$P(A \cup B) = P(B) + P(A \cap B^c) \qquad\qquad 4.4$$

Now substitute from Equation 4.3 for $P(A \cap B^c)$ into Equation 4.4 to obtain

$$P(A \cup B) = P(B) + P(A) - P(A \cap B)$$

which yields the desired Equation 4.1. This equation allows you to determine any one of the above quantities if you know the other three. For instance, if you know that A and B each have probability .5, and $A \cap B$ has probability 1/3, then it follows that $A \cup B$ has probability 2/3.

Problems 4.7

Caution: In these problems you not only have to worry about equation 4.1, but also about the rules of probability, Definition 4.6 on page 100. Please justify your answers.

1. In terms of the probabilities of A and B, what is the largest possible value of $P(A \cup B)$?

2. In terms of the probabilities of A and B, what is the smallest possible value of $P(A \cup B)$?

3. In terms of the probabilities of A and B, what are the smallest and largest possible values of $P(A \cap B)$?

Even with the bare-bones set of rules given in Definition 4.6, we can establish some *not-so-obvious* results, namely the second of the following Bonferroni inequalities.

Theorem 4.8: Bonferroni Inequalities

If $A_1, A_2,...$ is any sequence of events, then

$$P\left(\bigcup_{\text{all } i} A_i\right) \le \sum_{\text{all } i} P(A_i) \qquad 4.5$$

$$P\left(\bigcap_{\text{all } i} B_i\right) \ge 1 - \sum_{\text{all } i} (1 - P(B_i)) \qquad 4.6$$

❑

The first of these results is easy to believe, for the following reason. For any two events, A and B, whose probabilities are fixed, the proportion of outcomes in $A \cup B$ increases as you cut down the *size* of their overlap. So the maximum increase in $P(A \cup B)$ occurs when there is no overlap — in which case, from the rules of probability, we have $P(A \cup B) = P(A) + P(B)$.

Realize that there are not many tools available to provide a precise proof of these results, and the only tool which deals with the probability of a union, requires that the events forming this union be mutually exclusive. So about all we can do is to try to write $\bigcup_{\text{all } i} A_i$ as a union of mutually exclusive events. This will be done as soon as we establish the following very obvious result.

Lemma 4.9

If the event A is a subset[1] of the event B, then $P(A) \le P(B)$.

❑

This is easily established by noticing that $B = A \cup (B \cap A^c)$. This is a *disjoint union*, because its parenthesized element is a subset of A^c. Hence,

$$P(B) = P(A) + P(B \cap A^c) \ge P(A),$$

the inequality justified by the fact that probabilities are never negative.

Now to write $\bigcup_{\text{all } i} A_i$ as a union of mutually exclusive events, we simply replace A_2 by the part of A_2 that is not in A_1 — namely by $A_2 \cap A_1^c$ and in general, for each $n \ge 2$ replace A_n by the part of A_n outside of $\bigcup_{i < n} A_i$ — namely by

$$B_n = A_n \cap \left(\bigcup_{i < n} A_i\right)^c \qquad 4.7$$

If we define $B_1 = A_1$, then just from the way we constructed the B_n it should be evident that

$$\bigcup_{\text{all } i} A_i = \bigcup_{\text{all } i} B_i$$

---

1. A is a subset of B, written $A \subseteq B$, means that every element of A is also an element of B.

and that the B_i are mutually exclusive events. So

$$P\left(\bigcup_{\text{all } i} A_i\right) = P\left(\bigcup_{\text{all } i} B_i\right) = \sum_{\text{all } i} P(B_i) \qquad\qquad 4.8$$

But we can see from Equation 4.7 that for each n, B_n is a subset of A_n, from which it follows by Lemma 4.9 that $P(B_n) \le P(A_n)$. Substituting this inequality into Equations 4.8 yields

$$P\left(\bigcup_{\text{all } i} A_i\right) = \sum_{\text{all } i} P(B_i) \le \sum_{\text{all } i} P(A_i)$$

from which the first of the Bonferroni inequalities (Equation 4.5 on page 103) follows. For the second of these inequalities (Equation 4.6 on page 103) we can prove it from the first one by means of the second of the two DeMorgan's laws (Problems 4.5.1.d. on page 98, with t replaced by i)

$$\left(\bigcap_{\text{all } i} B_i\right)^c = \bigcup_{\text{all } i} B_i^c$$

by taking the probability of both sides of this equality, obtaining

$$P\left(\left(\bigcap_{\text{all } i} B_i\right)^c\right) = P\left(\bigcup_{\text{all } i} B_i^c\right)$$

and applying the Bonferroni inequality just proved to the right side of this equation, and the formula for the probability of the complement of any event to the left side, obtaining

$$1 - P\left(\bigcap_{\text{all } i} B_i\right) \le \sum_{\text{all } i} P(B_i^c)$$

or

$$1 - P\left(\bigcap_{\text{all } i} B_i\right) \le \sum_{\text{all } i} (1 - P(B_i))$$

which is seen to be equivalent to the second of the Bonferroni inequalities.

Now that we have the Bonferroni inequalities available, we see immediately, in a precise, quantitative fashion, that

- the **union** of a small number of *low-probability* events is still a *low-probability* event,

- the **intersection** of a small number of *high-probability* events is still a *high-probability* event.

The example that we now present illustrating Bonferroni's inequalities has that artificial flavor that often accompanies the need to provide immediate illustrations. In practice, meaningful applications of these inequalities occur frequently when we desire to show that the combination of several high performance statistical procedures may be thought of itself as a single high performance statistical procedure, no matter how these procedures may be related to each other.

Example 4.11 Probability of good weather

Suppose SW stands for sunny weather, PC stands for pleasantly cool, and G = $SW \cap PC$ stands for good weather. If $P(SW)$ = 0.995 and $P(PC)$ = 0.981 , then from the second Bonferroni inequality (Inequality 4.6 on page 103) It follows that

$$P(G) = P(SW \cap PC) \geq 1 - (P(SW^c) + P(PC^c))$$
$$= (1 - (0.005 + 0.019))$$
$$= 0.976.$$

So if the two components making up *good weather* are very likely, then *good weather* is still pretty likely.

∎

4.3 Discrete Uniform Models and Counting

When you have enough foresight (or luck) to have a uniform assignment of probabilities in a finite sample space, S, (i.e., all possible outcomes are equally likely), then, from the rules of probability (Definition 4.6 on page 100) for every event, A, it follows that

$$P(A) = \frac{\text{\# of possible outcomes in } A}{\text{\# of possible outcomes in } S}.$$ 4.9

So, if you think counting the number of elements of various sets is easy, you would put a great deal of effort into obtaining uniform models. Well, counting is not really as easy as it might seem, for reasons which will be soon explained. Nonetheless, there are a number of important cases in which being able to reduce a problem to one of counting does simplify the task of finding probabilities considerably.

- One of these is the computation of probabilities of events associated with random sampling — such as arise in quality control where we choose a sample of n objects at random from a set of size N, which has G good objects and $D = N - G$ defective ones. It is of interest to determine the probability that the sample contains g good units and $d = n - g$ defectives.

- Computing the probabilities of various events associated with card games such as bridge or poke is another example of using counting to compute probabilities under random sampling.

- Determination of the formula for the family of binomial distributions furnishes another example of the uses of counting.

Before going on to a systematic reliable approach to counting, we take a more careful look at what can make counting difficult. Generally, what we are trying to count is the number of objects in some given set which satisfy some given property. For instance

> • the number of subsets of size n from a set of size N 4.10

and

> • the number of subsets of size n having d defectives
>   from a set of size N which has D defectives.   4.11

Granted that it's quite easy to determine whether or not a subset of a set of size N is itself of size n , or whether a subset of size n does or does not have d defectives. If we could list all of the possible subsets, then by examining each one, we could find the number of sets in Expression 4.10; and by listing all subsets of size n , we could determine the number of subsets given by Expression 4.11. This method is aptly called the *method of exhaustion*, and tends to be totally impractical in most cases, even on a computer — because (as we will see) of the enormous number of subsets of a set consisting of a large number of elements.

To sum up, just knowing the properties which determine whether a given object is in some set hardly gives us any obvious immediate information about the size of this set.

So we now investigate how to improve on the method of exhaustion for counting the size of some sets.

Topic 4.10: Systematic counting methods

We start off informally with a simple example.

Example 4.12: Counting the number of H-T sequences

There are 2 such sequences of length 1, namely   (H) and (T)

There are 2×2 such sequences of length 2,
    2 arising from (H), namely (H,H) and (H,T)
    2 arising from (T), namely (T,H) and (T,T)

There are $2 \times 2 \times 2$ such sequences of length 3,
    2 sequences of length 3 for each of the 2×2 sequences of length 2.

It is evident that whenever another position is added (increasing the sequence length by 1) the number of sequences is doubled. So the number of H-T sequences of length n is 2^n .   ■

The example just presented points the way to a very useful generalization, whose proof is so similar to what was presented there that it will be left to the reader.

Theorem 4.11: Counting sequences

Suppose that we have a set, S, of sequences of length n, with the properties that

- the number of choices for position 1 of each sequence in S is the specified value c_1

- for each choice of specific elements in positions $1,...,k$ $(k < n)$ of sequences in S, the number of choices for the element in the $k+1$st position is the specified number c_{k+1}.

Then the number of sequences in S is the product $c_1 \times c_2 \times \cdots \times c_n$.

□

For H-T sequences, all c_i are 2.

For ordered samples of n elements with replacement from a set of size N, the value $c_k = N$ for all k.

For ordered samples of n elements without replacement from a set of size $N \geq n$, the value $c_k = N - (k-1)$.

Hence we have the following result.

Theorem 4.12: Corollary to Theorem 4.11, ordered sampling

The number of ordered samples of size k with replacement from a set of size N is N^k.

The number of ordered samples of size k without replacement from a set of size $N \geq k$ is $N \times (N-1) \times \cdots \times (N-[k-1])$.

□

At this point it is time to relate the number of ordered samples of size k without replacement from a set of size N, to the number of unordered samples of size k from a set of size N. This is easy to do because from Theorem 4.12 we see that $k \times (k-1) \times (k-2) \times \cdots 2 \times 1$ ordered samples of size k without replacement can be formed from an unordered sample of k distinct objects. Therefore

$$
\begin{aligned}
\frac{\text{\# of unordered samples of size } k}{\text{from a set of size } n} &= \frac{\text{\# of ordered samples of size } k \text{ from a set of size } n}{k \times (k-1) \times (k-2) \times \cdots 2 \times 1} \\
&= \frac{N \times (N-1) \times \cdots \times (N-[k-1])}{k \times (k-1) \times (k-2) \times \cdots 2 \times 1}
\end{aligned}
$$

4.12

We introduce the factorial notation to simplify dealing with Formula 4.12.

Definition 4.13: Symbols for j-factorial and the binomial coefficient

For each counting number $j = 1, 2, \ldots$ we let j-factorial be denoted by $j!$ defined by

$$j! = j \times (j-1) \times (j-2) \times \cdots 2 \times 1 \qquad\qquad 4.13$$

A more precise definition is the following.

$$0! = 1$$
$$\text{For each counting number } j \geq 0, \quad (j+1)! = (j+1) \times j! \qquad 4.14$$

The reason for defining $0! = 1$ is to make most mathematical formulas that arise using the factorials give the right answer. For instance, it makes the formula $(j+1)! = (j+1) \times j!$ valid for $j = 0$. This latter definition is preferable since it avoids the reader having to guess at the meaning of ... , while providing exact directions for how to compute $j!$ for any j. For instance
$$2! = 2 \times 1! = 2 \times 1 \times 0! = 2$$
$$3! = 3 \times 2! = 3 \times 2 \text{ etc.}$$

The binomial coefficient, denoted by $\binom{N}{k}$, defined for all integers, is 0 for N or k negative, as well as for $k > N$ is otherwise given by the formula

$$\binom{N}{k} = \frac{N!}{k! \times (N-k)!} \qquad\qquad 4.15$$

❑

Using the notation just introduced, we have the following.

Theorem 4.14: Corollary to Theorems 4.11 and 4.12

The number of unordered samples (subsets) of size k without replacement, which can be formed from a set of size N is the binomial coefficient (equation 4.15 above) $\binom{N}{k}$.

The number of ordered samples of size k (sequences of length k) without replacement, which can be formed from a set of size N is

$$k! \times \binom{N}{k} = \frac{N!}{(N-k)!} = N \times (N-1) \times \cdots \times (N - [k-1]) \qquad 4.16$$

❑

The formulas just presented are not very meaningful in isolation. Nobody really much cares that

$$\binom{100}{75} = 242519269720337121015504.$$

But these formulas are very useful for computing probabilities for card games and in theoretical investigations.

Example 4.13: Drawing from a deck of cards

Suppose we have a deck consisting of 52 distinct cards, half of which are colored red, and the other half colored black. If 13 cards are drawn at random what is the probability that exactly 5 of them are red?

Recalling Definition 3.2 on page 48, of random sampling without replacement, we find that the denominator of Equation 4.9 on page 105 is just the number of subsets of size 13 that can be formed from a set of size 52, which is, from Theorem 4.14 on page 108, the binomial coefficient $\binom{52}{13}$.

For the numerator of Equation 4.9, we need to determine the number of subsets of the 52 card deck which have 5 red and 8 black cards. In order to get the answer it seems easiest to represent each such subset as a sequence (S_r, S_b), where S_r is the subset of the 5 red cards, and S_b the subset of the 8 black cards. We can apply Theorem 4.11 on page 107, where $c_1 = \binom{26}{5}$ is the number of subsets of size 5 that can be formed from the set of 26 red cards, and $c_2 = \binom{26}{8}$ is the number of subsets of size 5 that can be formed from the set of 26 black cards. The desired probability of exactly 5 red cards is the ratio

$$\frac{\binom{26}{5}\binom{26}{8}}{\binom{52}{13}}$$

Using the definition of the binomial coefficient (Definition 4.13 on page 108) we find

$$\frac{\binom{26}{5}\binom{26}{8}}{\binom{52}{13}} = \frac{26! \times 26! \times 13! \times 39!}{5! \times 21! \times 8! \times 18! \times 52!}$$

$$= \frac{23 \times 19 \times 13 \times 13 \times 11 \times 11 \times 5}{49 \times 47 \times 43 \times 41 \times 17 \times 4}$$

$$= \frac{44681065}{276092852}$$

$$= 0.161833...$$

which gives the decimal value of the probability of 5 red and 8 black cards, when 13 cards are chosen at random from a deck with 26 black and 26 red cards.

∎

Example 4.14: Number of heads in fair coin tosses

In tossing a fair coin N times we want to determine the probability of exactly k heads. We assume that all head-tail sequences of length N have the same probability. Since the number of head-tail sequences of length N is 2^N (see Theorem 4.12 on page 107, where the N of this theorem is 2, and the k of this theorem is our N here) the probability of any specified such sequence must be $1/2^N$. From the basic rules for a probability measure (Definition 4.6 on page 100) to get the desired probability, we must multiply $1/2^N$ by the number of head-tail sequences of length N which have exactly k heads. But each sequence of k heads (and the remaining $N - k$ tails) is determined by the subset (of size k) of the set of the N positions having the k heads. Hence we must multiply by the number of subsets of size k which can be formed from a set of size N — and this is known (from Theorem 4.14 on page 108) to be the binomial coefficient $\binom{N}{k}$. So the probability of exactly k heads in N tosses of a fair coin is

$$\binom{N}{k}/2^N.$$

∎

Practice Exercises 4.15

1. What is an expression for the probability of:

    a. fewer than k heads in N tosses of a fair coin?

    b. less than or equal to k heads in N tosses of a fair coin?

    c. more than k heads in N tosses of a fair coin?

    d. at least k heads in N tosses of a fair coin?

2. In a lot of 100 digital computers, suppose that 18 are defective and the remaining ones are OK. If a sample of 20 computers is drawn at random from this lot.

    a. What is an expression for the probability that exactly 3 are defective?

    b. Find, to 3 digit accuracy, the probability that exactly 3 are defective. *Hints*: How different are these two questions from those investigated in Example 4.13 on page 109?

    c. Find, to 3 digits, the probability that fewer than 4 are defective.

3. A deck of cards for the game of *bridge* consists of 52 cards, with four equal size mutually exclusive subsets called *spades, hearts, diamonds* and *clubs*. A bridge *hand* is a subset of 13 of the 52 cards. If a bridge hand is chosen at random from the bridge deck, what is the probability that it consists of 3 spades, 2 hearts, 5 diamonds and 3 clubs? (Find both an expression for this probability, and its approximate decimal value.) *Hint*: Extend the reasoning of Example 4.13 on page 109.

4. A bridge deck (see Question 3. of this set) is dealt out to 4 people, each of whom receives a hand of 13 cards. Suppose it is known that two of the people (the *opposing partners*) received a total of 7 hearts. Assuming that their hands could be thought of as being drawn at random from a deck of 7 Hearts and 19 *non-hearts*, what is the probability of a 4-3 split? (i.e. that either one of the opposing partners gets exactly 4 hearts, the other getting 3 hearts)?

5. For those of you familiar with other card games, such as poker or pinochle, make up some questions that might be of interest to players of these games.

Problem 4.15

The so-called *Bose-Einstein* statistics concern the probability distribution of energy levels assigned to photons. Assuming n possible distinguishable energy levels, the distribution of energy levels to r electrons is assumed to be uniform over the set of distinguishable assignments to these indistinguishable photons. Determine the number of such distinguishable assignments — the probability of each such assignment being the reciprocal of this number.

Hint: Represent the n possible distinct energy levels by the spaces between $n + 1$ bars, as shown here for $n = 4$: $| \; | \; | \; | \; |$. Let r_i denote the number of photons in cell i, for $i = 1,...,n$. The number of distinguishable assignments is just the number of distinguishable sequences $(r_1,...,r_n)$ with

$$r_1 + ... + r_n = r.$$

Graphically each such assignment can be represented by a diagram like the one following: $|**|\,|***|*|$, where $r = 6$. Notice that there are $(n + 1) - 2 + r$ positions between the outside-boundary bars in which to put the energy level boundaries ($n - 1$ of them) and the photons (r of them). The number of such distinguishable assignments is just the number of possible choices of the subset of positions for the remaining $n - 1$ bars (or of the r photons) from the set of $n - 1 + r$ available positions.

Topic 4.16: A little more on counting

One of the difficulties in counting is that in many cases there is a good chance of double-counting. For instance, suppose you want to determine the probability that the sum on 5 successive rolls of a fair die is 27. If you try to approach this problem by listing all of the possible outcomes you are likely to leave out some of the outcomes, or just get exhausted and give up. Let's see how this problem might be tackled

Example 4.16: Probability of sum of 27 on 5 rolls of a fair die

What we do to deal with this problem is to look at the last roll, and use it to subdivide the event of interest (sum of 27) into 6 parts. So if the last roll resulted in the value 6, in order for any such outcome to result in a sum of 27, the first 5 rolls must result in a sum of 21, while if the last roll resulted in the value 5, the first 5 rolls must result in a sum of 22, etc. To get a handle on this process, we let $N_{k,n}$ denote the number of sequences of n die tosses which result in a sum of k. From the reasoning just given we find that

$$N_{k,n} = N_{k-1,n-1} + N_{k-2,n-1} + \cdots + N_{k-6,n-1}$$

which we write as

$$N_{k,n} = \sum_{j=1}^{6} N_{k-j,n-1}. \qquad 4.17$$

It is immediately evident that

$$N_{k,1} = \begin{cases} 1 & \text{if } k = 1, 2, 3, 4, 5, 6 \\ 0 & \text{otherwise.} \end{cases} \qquad 4.18$$

We can use equation 4.17, together with its *initial conditions,* equations 4.18, in three different ways.

- A computer can be programed to numerically determine successively all values of $N_{k,2}$ from those of $N_{k,1}$, for instance

$$N_{1,2} = N_{0,1} + N_{-1,1} + N_{-2,1} + N_{-3,1} + N_{-4,1} + N_{-5,1} = 0$$

$$N_{2,2} = N_{1,1} + N_{0,1} + N_{-1,1} + N_{-2,1} + N_{-3,1} + N_{-4,1} = 1$$

$$N_{3,2} = N_{2,1} + N_{1,1} + N_{0,1} + N_{-1,1} + N_{-2,1} + N_{-3,1} = 1 + 1 = 2$$

$$N_{4,2} = N_{3,1} + N_{2,1} + N_{1,1} + N_{0,1} + N_{-1,1} + N_{-2,1} = 1 + 1 + 1 = 3$$

  Once we have computed all values of $N_{k,2}$ we can repeat this process to compute all values of $N_{k,3}$, etc.

- If we have somehow guessed a formula for the $N_{k,n}$, it can be verified by showing that it satisfies Equations 4.17 and 4.18.

- There are analytic procedures (generating functions) which are sometimes useful for finding formulas for sequences satisfying *difference equations* like 4.17 with initial conditions like 4.18.

Once we have such formulas, there are frequently good approximations for them (such as those that can be obtained using the Stirling formula approximation for factorials).

The basic idea behind the approach which led to Equations 4.17 with initial conditions 4.18 is the following. If we want to determine the number of elements in some set, A, we try to break up this set into a disjoint, exhaustive union of subsets, say $A_1,...,A_m$, all of which have a structure which is either similar to the original set, A or related in a simple fashion to such a subset. This is illustrated in Figure 4-4. Repeat this process on each of the $A_1,...,A_{m_1}$, generating $A_{1,1},...,A_{1,m_2}$, etc. Repeat this process until you get down to sets small enough to count.

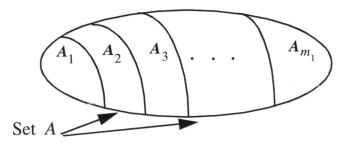

Figure 4-4 Subdivision of Set A into Smaller, Similar Structured Subsets ∎

Problems 4.17

1. Derive a difference equation with initial conditions for the number of subsets of size k of a set of size n. Compute for $n = 4$ and all k.
   Hint: Subdivide into those subsets which contain a particular element and those subsets which do not contain this element.

2. Derive a difference equation with initial conditions for the number of head-tail sequences of length n having exactly k heads. From these, compute the number of such sequences of length 4 having exactly 2 heads.

3. Derive a difference equation with initial conditions for the number of H-T sequences of length n having a run of at least k successive heads.
   Hint: Either the last toss completes the first such sequence, or the first such sequence has been completed already.

L

4.4 Conditional Probability

When additional information becomes available, we have to know how to take account of it; that is, it is essential that the model be modified appropriately. During an election, information about how the voting is going causes the pollsters to update their predictions continually. The government frequently recalculates the inflation rate and the cost of living, in light of the latest data. In space flight, course corrections are made whenever available measurements indicate them. Hurricane probability predictions are modified with information coming in from satellites and hurricane information flights.

So how do we take account of such incoming information? First we need a precise formulation of what is meant by *new information*. Experience has shown that for the purposes of updating a probability model, the following is appropriate. Suppose it is guaranteed that whenever an experiment is performed, the occurrence or nonoccurrence of an event B will be known. Under these circumstances, we would like to know how we should modify the probability assigned to the event A when B has occurred.

The simplest reasoning states that under the given circumstances the event B should be considered as the updated sample space. Thinking of the probability of A as the *proportion that A takes up in the sample space*, the proportion that A takes up in the *new sample space*, B, is

$$\frac{P(A \cap B)}{P(B)}$$

as indicated in the Venn diagram of (B is indicated by the horizontal hatching).

Figure 4-5 The Event B is the Updated Sample Space

The frequency interpretation of probability also supports this definition, because the long-run proportion of occurrences of the event A, among the experiments in which B occurred is

$$\frac{\text{number of the } n \text{ experiments in which } A \cap B \text{ occurred}}{\text{number of the } n \text{ experiments in which } B \text{ occurred}}$$

which equals

$$\frac{(\text{number of the } n \text{ experiments in which } A \cap B \text{ occurred})/n}{(\text{number of the } n \text{ experiments in which } B \text{ occurred})/n}$$

which gets close to the ratio $\dfrac{P(A \cap B)}{P(B)}$ for large n, provided $P(B) \neq 0$.

These arguments lead us to the following.

Definition 4.18: Conditional probability of A given B

The conditional probability of the event A given (the occurrence of) the event B, denoted by $P(A|B)$, is defined by the equation

$$P(A|B) = \frac{P(A \cap B)}{P(B)} \qquad\qquad 4.19$$

provided $P(B) \neq 0$.

It is interpreted as the long run proportion of occurrences of the event A among all experiments in which the event B occurs.

❑

Example 4.17: $P(7 \text{ or } 11| {>}4)$

Tossing a pair of fair dice, if B is the event that the sum on the dice exceeds 4, what is the probability of A, that the sum will be either 7 or 11?

We know that $P(A) = 1/6$ (see Examples 1.1 on page 8 and 4.9 on page 96). We surely expect that $P(A|B) > 1/6$. Let's see.

We also know that

$$P(B) = 1 - P(B^c) = 1 - P\{(1, 1), (1, 2), (2, 1), (3, 1), (1, 3), (2, 2)\}$$
$$= 5/6$$

Since both 7 and 11 exceed 4, we see that $A \cap B = A$. Hence

$$P(A|B) = \frac{P(A \cap B)}{P(B)} = \frac{6}{5}P(A) = \frac{6}{5} \times \frac{1}{6} = \frac{1}{5}.$$

■

Many models are specified by assigning both conditional and ordinary (unconditional) probabilities. For instance, random sampling without replacement can be accomplished by random sampling one at a time from the objects remaining in the set sampled. That is, the conditional distribution of the j-th object drawn is uniform over the objects which have not yet been drawn. Again this appears easier to understand than the description stating that random sampling is defined by having all subsets of a given size equally likely.

In physics, one of the most meaningful ways to present the so-called *Fermi-Dirac statistics*, which describe the distribution of electron energy levels, is by giving the conditional probabilities that a randomly assigned electron will *choose* a specified energy level, given the energy levels of the electrons already present, and specifying the (unconditional) energy level distribution of the first electron. Supposing a finite number, n, of possible energy levels, labeled $1,2,...,n$, with any given energy level capable of being occupied by at most one electron, the usual formulation for the Fermi-Dirac model is that for $r \leq n$ electrons, all possible distinguishable such assignments are equally likely. The number of such possible assignments of the indistinguishable electrons to the distinguishable energy levels is the binomial coefficient $\binom{n}{r}$, since each assignment is determined by the set of energy levels occupied by the r electrons. Considering the electrons to be assigned energy levels one at a time, it can be shown that we get the same model by assuming the energy level of the first electron to be chosen at random from the set of n energy levels, and that given the j energy levels occupied by the first j electrons, the conditional probability distribution of the $j+1$st electron is uniform on the set of n - j unoccupied energy levels. Again this description seems simpler to understand than the originally given one.

When we talk about the risk of acquiring AIDS among various populations, we are referring to conditional probabilities of getting AIDS, given that an individual was randomly chosen from one of the various populations.

Practice Exercises 4.18

1. In tossing a pair of dice, what is the conditional probability of a sum of 7, given that the sum was either 7 or 11? (Assume a uniform distribution on the sample space of 36 ordered pairs (see Example 1.1 on page 8)

2. Define a sample space in which you can meaningfully speak of the conditional probability of getting lung cancer, given that you smoke.

3. Suppose that 5 success-failure experiments are run, and that each sequence in the sample space has probability $1/2^5$. What is the probability of exactly k successes given no failures, for $k = 0,1,...,5$?

4. Let T stand for the lifetime of a standard 100 watt light bulb, and suppose that $P(T < t) = 1 - e^{-t/2500}$ for $t > 0$.

   a. What is the (theoretical) median lifetime of such a bulb?

   b. What is the probability that the bulb will last for less than t more hours, given that it has already lasted for τ hours?

5. In tossing a pair of fair dice (assuming a uniform distribution on the sample space of 36 ordered pairs (see Example 1.1 on page 8), what is the probability of a 3 on the second toss, given a 4 on the first one?

6. Supposing a uniform distribution on the sample space of 5 repeated 50-50 experiments.

   a. What is the probability of a total of 5 successes, given that the first trial was a success?

   b. What is the probability of a total of 5 successes, given that at least one trial was successful?

A very fundamental result involving conditional probability is the *law of total probability* (sometimes called the *stratified sampling theorem*). The idea behind this result is a simple one, namely that if you know the proportion of occurrences of some event, *A*, in each of a set of mutually exclusive, exhaustive subpopulations, and we know the size of each of these subpopulations, then we can compute the overall proportion of occurrences of *A* in the entire population.

Problem 4.19

Determine how to carry out the computation just referred to.

Theorem 4.20: The stratified sampling theorem

Suppose that the events $B_1, B_2,...$ are mutually exclusive ($B_i \cap B_j = \phi$ when $i \neq j$) and exhaustive

$$\bigcup_{\text{all } j} B_j = S$$

(S is the sample space, see Definition 4.1 on page 94) then

$$P(A) = \sum_{\text{all } j} P(A|B_j)P(B_j) \qquad 4.20$$

◻

To establish this result look at the following computations.

$$P(A) = P(A \cap S) = P\left(A \cap \bigcup_{\text{all } j} B_j\right)$$

$$= P\left(\bigcup_{\text{all } j}(A \cap B_j)\right) = \sum_{\text{all } j} P(A \cap B_j)$$

$$= \sum_{\text{all } j} P(A|B_j)P(B_j)$$

Figure 4-6 might help in following some of steps above.

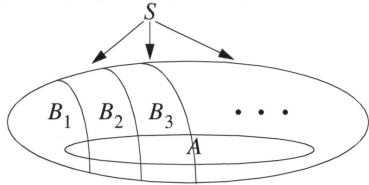

Figure 4-6 The Geometry Behind the Stratified Sampling Theorem

We illustrate this theorem with what is called an *urn model*. Although such models are presented as problems arising from drawing labeled balls from urns, they can be used to model a variety of real problems, such as spread of infectious diseases, and spread of rumors and the behavior of pollution.

Example 4.19: Simple urn model

An urn contains 5 red balls and 25 black ones. A ball is chosen at random from the urn. If it is red, it and 3 additional red balls are returned to the urn. If it is black, it and 1 additional black balls are returned to the urn. Immediately after the completion of drawing and returning balls to the urn, what is the probability that the second drawing results in a red ball?

To solve this, we let R_j be the event that the j-th drawn ball is a red one. We want to calculate $P(R_2)$. We use the stratified sampling theorem (Theorem 4.20) as follows.

$$P(R_2) = P(R_2|R_1)P(R_1) + P(R_2|R_1^c)P(R_1^c)$$

$$= \frac{8}{33}\frac{5}{30} + \frac{5}{31}\frac{25}{30} = \frac{1073}{6138}$$

$$\cong 0.1748$$

■

Practice Exercises 4.20

1. Referring to Example 4.19, calculate $P(R_3)$ and $P(R_1|R_2)$.

2. A pair of 3-sided fair dice are tossed (possible results, 1, 2, and 3).

    a. What is the probability of an even sum?

    b. What is the probability that the dice show the same face, given that the sum is odd?   Given that the sum is even?

3. Two people are chosen at random without replacement (in order) from a population of 15 men and 10 women. Given that the second person chosen is a woman, what is the probability that the first one chosen is a woman.

4. Suppose that 20% of rich families, 70% of middle class families and 10% of poor families own 3 TVs; and that 5% of families are rich, 60% of families are middle class and the remaining families are poor. What proportion of families own 3 TVs?

Problem 4.21 The three drawer problem

A cabinet has 3 drawers. The first drawer has two distinct gold coins. The second drawer has a silver and a gold coin. The third drawer has 2 distinct silver coins. A drawer is chosen at random, and from it a coin is drawn at random. What is the probability that the other coin in the chosen drawer is silver, given that the chosen coin is gold?

Hint: Choose an appropriate sample space (whose elements might be of the form $((D_i, C_{j,i})$ $i = 1,2,3, j = 1,2$ $C_{j,i}$ designates the coin (1-st or 2-nd) chosen from the i-th drawer). Now define appropriate events in **this sample space** (e.g., *chosen coin is gold, other coin is gold,* etc.) and assign probabilities and conditional probabilities from the information supplied.

⌈

An interesting application of the stratified sampling theorem arises in medical diagnosis. Here is it assumed that we know the prevalence $P(C_i)$, $i = 1,2,...,I$, of a sequence of mutually exclusive and exhaustive health conditions (many of which will be diseases), as well as the conditional probabilities, $P(S_j|C_i)$, of *symptom sets, S_j, $j = 1,2,...,J$*, given each of the health conditions C_i. For instance one of the health conditions might be *measles,* and the most likely symptom set might be {fever, rash of a given type, headache,...}. With the given information, Bayes' theorem states that

$$P(C_i|S_j) = \frac{P(S_j|C_i)P(C_i)}{\sum\limits_{k=1}^{I} P(S_j|C_k)P(C_k)} \qquad\qquad 4.21$$

This is easily established by using the definition of conditional probability in the numerator, and the stratified sampling theorem in the denominator as shown.

$$P(C_i|S_j) = \frac{P(C_i \cap S_j)}{P(S_j)} = \frac{P(S_j|C_i)P(C_i)}{\sum\limits_{k=1}^{I} P(S_j \cap C_k)}$$

This allows the diagnostician to determine which state of health is most likely, from various disease prevalences and conditional symptom set probabilities.

Bayes theorem is usable in statistics problems, where we have some prior knowledge that the unknown parameter(s) (such as the probability of success in some experiment) have a known probability distribution.

L

Topic 4.22: Relation between random sample and random sampling one at a time without replacement

In order to see the relation between sampling *one at a time* without replacement and drawing a random sample without replacement (Definition 3.2 on page 48), the following theorem is useful.

Theorem 4.23: Probability of intersection and conditional probability

Let B_i, $1 \le i \le k$ be any sequence of at least two events. Then

$$P\left\{ \bigcap_{i=1}^{k} B_i \right\} = P\{B_1\} \prod_{i=2}^{k} P\left\{ B_i \Big| \bigcap_{j<k} B_j \right\} \qquad 4.22$$

For $k = 3$ this reads

$$P(B_1 \cap B_2 \cap B_3) = P\{B_1\} P\{B_2 | B_1\} P\{B_3 | B_2 \cap B_1\}$$

❏

Strictly speaking, the proof is by induction. The induction is so easy that one may as well be informal, and substitute for the conditional probabilities on the right, and noticing the *telescoping cancellation* in this finite product.

If we now sample j times, one at a time without replacement at random from a given set of n objects, the probability of getting any one particular sequence (which must consist of distinct elements, because of the nature of the sampling) is, by Theorem 4.23,

$$n \times (n-1) \times \ \cdots \times (n - [j-1]) \, .$$

The probability of the event consisting of all sequences (of distinct elements) whose elements are specified is the above probability, multiplied by the number of such sequences. This latter number is easy to find, because from Theorem 4.12 on page 107 we see that $j \times (j-1) \times \cdots \times 2 \times 1 \ = \ j!$ ordered samples of size j without replacement can be formed from any unordered sample of j distinct objects.

So the probability of the event that the set of objects drawn without replacement consists of j distinct chosen objects is the following:

$$\frac{j \times (j-1) \times (j-2) \times \cdots \times 2 \times 1}{n \times (n-1) \times \ \cdots \times (n - [j-1])} = \frac{1}{\binom{n}{j}}.$$

But this latter number is just the probability of any particular set of j elements drawn at random from a set consisting of n elements; because there are $\binom{n}{j}$ such sets of size j, they are all equally likely (definition of random sample) and their sum must be 1. So each one must have probability $1/\binom{n}{j}$, as asserted. This establishes the assertion that a random sample without replacement can be drawn by drawing one object at a time at random without replacement.

4.5 Statistical Independence

Suppose that we have two success-failure experiments.   If we claim that the two experiments are (statistically) independent of each other, we should mean that we don't change our probability assignments for either experiment even knowing the results of the other one. If we let S_1 be the event *success on the first experiment*, and S_2 be the event *success on the second experiment.* this would certainly mean that

$$P(S_1|S_2) = P(S_1), \quad P(S_1^c|S_2) = P(S_1^c)$$

$$P(S_2|S_1) = P(S_2), \quad P(S_2^c|S_1) = P(S_2^c)$$

If on the other hand, you claimed, in choosing a person at random from our population, that (the events) *person smokes* (S) and *person gets lung cancer* (L) are dependent, then we would expect that at least one of the following inequalities holds:

$$P(L|S) \neq P(L), P(L^c|S) \neq P(L^c)$$

$$P(L|S^c) \neq P(L), P(L^c|S^c) \neq P(L^c)$$

(with the tobacco companies asserting that $P(L|S) < P(L)$).

It would seem reasonable to define statistical independence of any two events, A and B by

$$P(A|B) = P(A), \quad P(A^c|B) = P(A^c)$$

$$P(B|A) = P(B), \quad P(B^c|A) = P(B^c).$$

A small amount of algebra will lead to a simpler formulation. If, in fact, $P(A|B) = P(A)$, then it follows from the Definition 4.18 on page 114, of the conditional probability $P(A|B)$, that

$$\frac{P(A \cap B)}{P(B)} = P(A)$$

from which it follows that

$$P(A \cap B) = P(A)P(B). \qquad\qquad 4.23$$

If $P(B) \neq 0$, it is evident that when Equation 4.23 holds, then $P(A|B) = P(A)$. Hence we use this equation as our definition of statistical independence.

Definition 4.24: Statistical independence of events A and B

Events A and B are said to be *statistically independent* if

$$P(A \cap B) = P(A)P(B)$$

$\square$

It might seem reasonable to define statistical independence of any sequence of events, $A_1, A_2, A_3, \ldots$ as the statistical independence of every distinct pair of events from this sequence. What follows shows that this does not work as we might expect.

Example 4.21: Dependence of three pairwise independent events

In the toss of a pair of fair dice let

$E1$ be the event *even on first die*, $E2$ be the event *even on second die* and ES be the event *even sum*.

Using the uniform model on the 36 point sample space of Table 1.1 on page 8, it can be verified that all pairs from these three events are statistically independent, each having probability 1/2. However the three are not mutually independent, if we want this to mean that every event formed from every two of them is independent of every event formed from the remaining event; because whenever both $E1$ and $E2$ both occur, ES occurs, and hence $(E1 \cap E2) \cap ES = E1 \cap E2$. Thus

$$\frac{1}{4} = P(E1 \cap E2) = P((E1 \cap E2) \cap ES) \neq P(E1 \cap E2)P(ES) = \frac{1}{8}$$

showing that $E1 \cap E2$ and ES are not independent of each other. That is, even though $E1, E2$ and ES are pairwise independent, they are not mutually independent. ∎

So how are we supposed to find the simplest definition of mutual independence of a sequence of events? The answer which has the desired properties (that any event formed from any subsequence via set operations is statistically independent of any other event formed from the remaining elements of the sequence, is given in the following definition. The proof that this definition gives the desired results is beyond the scope of this text

Definition 4.25: Mutual statistical independence

We define the mutual statistical independence of the events in the sequence A_i, $i = 1, 2, \ldots$ as follows. Let A_i^* stand for any of the events A_i, A_i^c, S, ϕ. The events A_i, $i = 1, 2, \ldots$ are mutually statistically independent if for each subset, I of the set C, of counting numbers for which A_i, $i = 1, 2, \ldots$ are defined, the events

$$\bigcap_{j \text{ in } I} A_j^* \quad \text{and} \quad \bigcap_{k \text{ not in } I} A_k^*$$

are statistically independent; that is, if

$$P\left(\left(\bigcap_{j \text{ in } I} A_j^*\right) \cap \left(\bigcap_{k \text{ not in } I} A_k^*\right)\right) = P\left(\bigcap_{j \text{ in } I} A_j^*\right) P\left(\bigcap_{k \text{ not in } I} A_k^*\right)$$

☐

Note that just requiring that

$$P\left(\bigcap_{j \text{ in } C} A_j\right) = \prod_{j \text{ in } C} P(A_j)$$

(where the right hand side of this equation stand for the product of the $P(A_j)$) won't accomplish what is wanted, since this equation will hold if A_1 is the empty set, ϕ.

It is helpful to realize that it is rare that we have to verify the statistical independence of sequences of events mathematically.[1] What we usually do, is to take pains to ensure that certain chosen events are statistically independent —in the sense that knowledge of the occurrence or nonoccurrence of some of these events should not affect the chance of occurrence or nonoccurrence of the remaining ones. So in clinical trials comparing two different treatments for the same disease, we try to ensure both random assignment of the treatments and that neither patients nor individuals involved with administering the treatment (doctors, nurses, or patients) know which treatments are given to which people. In comparing different industrial processes, care must be taken that the results for one process don't cause changes in the remaining processes. So, for example, if the annealed steel in the first run is too hard, we don't adjust the temperature in the second run on this account. When these conditions are satisfied, in our mathematical analyses we can legitimately treat these events as statistically independent — often making the analysis feasible where it might not have been, or much simpler instead of very complicated.

Example 4.22: $P(A \cup B^C)$

Suppose A and B are statistically independent with $P(A) = 0.8$ and $P(B) = 0.95$. We want to determine $P(A \cup B^C)$. One way to do this is to use equation 4.1 on page 101, which yields

$$\begin{aligned}
P(A \cup B^C) &= P(A) + P(B^C) - P(A \cap B^C) \\
&= 0.8 + (1 - 0.95) - 0.8(1 - 0.95) \\
&= 0.81.
\end{aligned}$$

∎

A preferable solution to this problem involves use of one of the DeMorgan Laws (Problems 4.5.1.d. on page 98).

---

1. In certain derivations, such as that of the student t distribution, it is necessary to verify that two quantities (in the t case, the numerator and denominator of the t statistic) are statistically independent.

Practice Exercises 4.23

1. Three independent success-failure experiments, each having probability, p, of success, are run. Given that there were exactly 2 failures, what is the probability that the first experiment was a failure?

2. Three independent success-failure experiments are run, with the first experiment having probability .5 of success, and the remaining ones having probability .25 of success. Given that there were exactly two successes, what is the probability that the first experiment was a failure?

3. Suppose that A, B, and C have probabilities .6, .7, and .8 respectively. Determine whether or not they are independent events.

4. Is it possible for $P(A \cup B \cup C) = 1$, if A, B, C are statistically independent events, none of which has probability 1?

5. Show that the model of Example 1.1 on page 8, for tossing a pair of fair dice, is the same as would result from independent tosses of a pair of fair dice.

Now that we have introduced a set of tools related to probability, it is helpful to present a semiformal approach to attacking problems in this field.

4.6 Systematic Approach to Probability Problems

1. State the problem in general terms, specifying the event whose probability is of interest.
2. Choose a sample space, S, in which
   a. The event of interest corresponds to a subset of S.
   b. All events with known probabilities (or conditional probabilities) correspond to subsets (or pairs of subsets) of S.
3. Assign the known probabilities and conditional probabilities.
4. Specify the relationships
   a. between the events with known probabilities and the event whose probability is of interest,
   b. between the corresponding probabilities.
5. Compute either a formula for the probability of interest or an algorithm for its computation; and/or a decimal approximation or bounds for this probability.

Here are a couple of examples of this process.

Example 4.24: Same face on three die tosses

1. We want to find the probability that the same face shows up on three independent tosses of a fair die.

2. A reasonable sample space, S, is the collection of all sequences (t_1, t_2, t_3) where the t_i can be any integer from 1 to 6; t_i corresponds to the face showing on toss i.

3. The events $A_{i,j}$ that the i-th toss comes up j, $i = 1,2,3$ $j = 1,...,6$ are all assumed to have probability 1/6 (the fairness of the die), and for $i \neq i^*$ and all j,j' $A_{i,j}$ and $A_{i^*,j'}$ are statistically independent.

(Note that each of these events consists of 36 elements of the sample space.)

4. From this it follows that each elementary event (subset of S consisting of a single point in S) is an intersection of three independent events, $A_{1,j_1} \cap A_{2,j_2} \cap A_{3,j_3}$, so that its probability is

$1/6^3$. The event of interest is the union $\bigcup_{1 \leq j \leq 6} A_{1,j} \cap A_{2,j} \cap A_{3,j}$

of six mutually exclusive events, each with probability $1/6^3$.

5. Hence the desired probability is $6/6^3 = \dfrac{1}{36} \cong 0.02778$.

∎

Example 4.25: Probability of a 4-3 split

1. Suppose a set of 26 playing cards, consisting of 6 *spades* and 20 *non-spades*, is randomly divided into 2 *hands* of 13 cards each. Find the probability of a 4-3 split of the spades — i.e., that either one of the two hands has 4 spades, and the other hand has 3 spades.

2. A good choice for the sample space, S, is the set of all subsets of 13 objects from the 26 available ones (the sample space consisting of the first of the hands drawn; the second hand is then completely determined). The event of interest consists of those subsets of the sample space which have either 4 spades (and 9 non-spades) or 3 spades (and 10 non-spades). For the purposes of dealing with this type of event, it is convenient to write each element of the sample space in the form of a sequence, ({k spades},{13 - k non-spades}) for $k = 3,4$. The event of interest consists of the above sequences with $k = 3,4$.

3. From Theorem 4.14 on page 108, we know that the number of elements of the sample space is the binomial coefficient $\binom{26}{13}$. Thus the probability of any single possible outcome is $1/\binom{26}{13}$.

4. From Theorem 4.14 on page 108, the number of subsets consisting of k spades which can be formed from the set of 6 available spades is the binomial coefficient $\binom{6}{k}$ and the number of subsets consisting of 13 - k *non-spades* is the binomial coefficient $\binom{20}{20-k}$. Thus, from Theorem

4.11 on page 107, the number of outcomes in the event of interest is

$$\binom{6}{3}\binom{20}{13-3} + \binom{6}{4}\binom{20}{13-4}.$$

Hence, due to the random sampling assumption, the probability of a 4-3 split is the ratio

$$\frac{\binom{6}{3}\binom{20}{10} + \binom{6}{4}\binom{20}{9}}{\binom{26}{13}}.$$

5. Each term, T, in the above ratio can be hand computed by writing

$$T = Exp(Log(T)),$$

where Log is the natural logarithm function. Doing this yields the result that the given 4-3 split has probability approximately 0.597517.

∎

4.7 Random Variables, Expectation and Variance

For many experiments the sample space does not consist of numbers. Toss a coin many times and the elements of the sample space are H-T sequences. In the experiment of obtaining an electrocardiogram, the sample space elements are usually sequences of functions of time, represented as graphs. Sample spaces like these can be fairly complicated, in order that they be able to give a fairly complete description of the experiments being made. Often we find it worthwhile to simplify the analysis by examining specific aspects of an experiment. The most common such simplification is the association of a number with each experimental outcome. As most of you know, such an association is called a function. When the inputs to such a function are experimental outcomes, and the outputs are numbers, under specific, so-called measurability conditions, such functions are called *random variables*.

Definition 4.26: Random variable

A real valued function, X, whose allowable inputs are elements of a sample space having a probability measure, and satisfying the following *measurability condition*

> that for each interval, I, the set of elements, s, in the sample space for which $X(s)$ belongs to I is an event in the sample space (i.e., has an associated probability)

is called a *random variable*. We use the abbreviation *rv* for a random variable.

❑

If you think of a random variable as converting from the original sample space, to a sample space consisting of real numbers, then the measurability condition is simply the requirement that in this new sample space, every interval must be an event (have a probability). A few illustrations of random variables are:

In tossing a coin n times, the random variable S_n associates with a coin toss sequence, the number of heads in the sequence. We just call S_n *the number of heads in n tosses*. Notice that S_n is not a number, but rather **a rule for obtaining a number**.

Another bunch of random variables associated with coin toss sequences are the X_i — the number of heads on the i-th toss (0 or 1). Then

$$\sum_{i=1}^{n} X_i = S_n .$$ $\qquad$ 4.24

Several different random variables that might usefully be associated with a single graph from an electrocardiogram are

the *total area* between the graph and the time axis
the *maximum height* of the graph
the *minimum height* of the graph
indicator functions to specify the presence (1)/absence (0) of
characteristics indicating various types of heart damage.

There are three main types of random variables which arise in applications:

- *Discrete random variables* which can take on only a finite number of values (for example, binomial random variables, see Topic 2.5 on page 30),

- *Discrete random variables* which can take on a countably infinite set of values (e.g., a value for each non-negative integer, for example Poisson variables, see Topic 2.6 on page 31),

- *Random variables with a probability density* — for example, Gaussian (*normal*) random variables (see Normal (Gaussian) Distributions, Topic 2.2 on page 24).

The probability distributions of the first two types of random variables can be specified by a discrete probability function, defined as follows;

Definition 4.27: Probability function of a discrete RV

If the set of possible values of the discrete random variable, X, is denoted by PV, (where PV is in one-to-one correspondence with a subset of the set of non-negative integers), then the discrete probability function, p, describing the probability behavior of X has as its value at v in PV, the quantity

$$p_v = P\{X = v\} ,$$

where the expression $\{X = v\}$ stands for the set of all those possible **experimental outcomes**, s, for which $X(s) = v$ (remember that the probabilities were associated with events in the sample space).

❑

Definition 4.28: Probability density of an RV

Similar to Definition 4.27, when X has a density, denoted by f_X, the expression $a \leq X \leq b$ stands for the set of all those possible **experimental outcomes**, s, for which $a \leq X(s) \leq b$,[1] and $P\{a \leq X \leq b\}$ is computed as the integral

$$\int_a^b f_X(x)\, dx$$

which we interpret geometrically as the area between the interval $[a;b]$ and the part of the graph of f_X above this interval. Most statistical computer packages have provisions to compute this area for the commonly encountered probability densities.

❏

We need to extend the concept of statistical independence to random variables, in order to talk meaningfully about population mean values.

Definition 4.29: Statistically independent random variables

Random variables $X_1, X_2,...,X_n$ are said to be statistically independent[2] if every event that can be formed using any subsequence $X_{i_1},...,X_{i_j}$ is statistically independent of every event that can be formed from the remaining X_i.

❏

We would like the population mean value of a random variable, X, to be the number that the value of the sample mean,

$$\frac{X_1 + X_2 +...+ X_n}{n}$$

4.25

---

1. If you want to think of some set of real numbers as the current sample space, then you may think of $P\{X = v\}$ as the probability of the event whose only element is the possible outcome v, and $P\{a \leq X \leq b\}$ as the event whose outcomes are the interval $[a: b]$.
2. The main purpose of the definition being given is to provide a sound theoretical basis for motivation of the definition of population mean. It is not meant as a practical means for verifying statistical independence of random variables — the following result being more useful in this regard.

   The random variables $X_1,...,X_n$ are statistically independent if and only if for all real numbers $x_1,...,x_n$, $P(\bigcap_{i=1}^{n} \{X_i \leq x_i\}) = \prod_{i=1}^{n} P\{X_i \leq x_i\}$.

tends to get close to as n grows arbitrarily large, where the X_i are independent random variables with the same probability distribution as X. This sequence of random variables may be thought of as resulting from independent repetitions of the experiment associated with X. These rvs are referred to as *independent, identically distributed rvs* (iid rvs).

To decide on what Expression 4.25 gets close to for large n, we first examine the case where X is a discrete random variable which can only take on a finite number of values, denoted by $x_1, x_2,...,x_k$. Examining the sum, $X_1 + X_2 +...+ X_n$, if the frequency interpretation of probability holds (see page 19) then for large enough n,

about np_1 of the $X_1, X_2,...,X_n$ take on the value x_1.

Their contribution to the sum $X_1 + X_2 +...+ X_n$ is thus about np_1x_1.

about np_2 of the $X_1, X_2,...,X_n$ take on the value x_2.

Their contribution to the sum $X_1 + X_2 +...+ X_n$ is thus about np_2x_2.

.

.

.

about np_k of the $X_1, X_2,...,X_n$ take on the value x_k.

Their contribution to the sum $X_1 + X_2 +...+ X_n$ is thus about np_kx_k.

Thus, from the frequency interpretation of probability, we expect the sum $X_1 + X_2 +...+ X_n$ to tend to be close to

$$np_1x_1 + np_2x_2 +...+ np_kx_k = n\sum_{j=1}^{k} p_jx_j$$

and thus for large n we expect the average

$$\frac{X_1 + X_2 +...+ X_n}{n}$$

to be close to the sum $\sum_{j=1}^{k} p_jx_j$. This motivates the following definition.

Definition 4.30: Population mean, mathematical expectation

Let X be a random variable which can take on a finite number of values,

x_1 with probability p_1

x_2 with probability p_2

$$\vdots$$

x_k with probability p_k

The mathematical expectation (or expectation) of X (which can also be thought of as the mean of the numerical population determined by X) denoted by $E(X)$ is defined as

$$E(X) = \sum_{j=1}^{k} p_j x_j \qquad 4.26$$

The meaning of the mathematical expectation, $E(X)$, is the value that the average

$$\frac{1}{n} \sum_{i=1}^{n} X_i = \frac{X_1 + X_2 + \ldots + X_n}{n}$$

tends to approach for large n, where $X_1, X_2, \ldots, X_n$ are independent random variables with the same probability distribution as X.

❑

For those with a background in infinite series and advanced calculus, we extend the definition of mathematical expectation. Those without this background will lose little by pretending that an infinite series is a finite sum. As for integrals, they may be thought of as areas.

Definition 4.31: Mathematical expectation in the countably infinite case

If X is a random variable with $P\{X = x_i\} = p_i$ for each positive integer, i, where

$$\sum_{i=1}^{\infty} p_i = 1$$

the mathematical expectation $E(X)$ is defined by the equation

$$E(X) = \sum_{i=1}^{\infty} x_i p_i \qquad 4.27$$

provided that the infinite series $\displaystyle\sum_{i=1}^{\infty} |x_i| p_i$ converges[1].

❑

To motivate the definition of mathematical expectation in the density case, just notice that if the density, f_X has a reasonably well-behaved graph, then the product $f_X(x)\Delta x$ for small Δx is approximately the area under the graph of f_X over the interval from $x - \Delta x/2$ to $x + \Delta x/2$ — i.e. $f_X(x)\Delta x$ **is approximately the probability of an outcome being in the interval from** $x - \Delta x/2$ to $x + \Delta x/2$. Since mathematical expectation is a sum of terms of the form value × probability of that value, the expression $x f_X(x)\Delta x$ is the equivalent of the term $x_i p_i$. A sum $\sum_i x_i f_X(x_i)\Delta x$, where

---

1. See Rosenblatt and Bell, [9], for a background on infinite series.

the x_i are a distance Δx apart, is an approximation to the area under the curve whose formula at x is $xf_X(x)$. This should be enough to motivate the following definition.

Definition 4.32: Mathematical expectation in the density case

If X is a random variable with a density, then the *mathematical expectation of X* is defined by the equation

$$\int_{-\infty}^{\infty} xf_X(x)\,dx \tag{4.28}$$

provided that $\int_{-\infty}^{\infty} |x|f_X(x)\,dx$ (which is the value approached by $\int_{-n}^{n} |x|f_X(x)\,dx$ as n gets arbitrarily large) is finite.[1]

In all of the above cases, the interpretation of $E(X)$ as the value approached by the sample average, $\frac{1}{n}\sum_{i=1}^{n} X_i$ as n gets large, remains valid.

❑

Example 4.26: Bernoulli expectation

An experiment with probability p of success and $1 - p$ of failure is called a *Bernoulli trial*. The random variable X, whose value is the number of successes on this trial, is called a Bernoulli variable with probability p of 1. Here $P\{X = 1\} = p, P\{X = 0\} = 1 - p$. The mathematical expectation of X, also called the *expected value of X*, is given by

$$E(X) = \sum_{i=0,1} x_i p_i = 1 \times p + 0 \times (1 - p) = p.$$

That is, the expected number of successes on 1 trial is p, (the probability of success per trial).

■

Practice Exercises 4.27

1. What is the expected value of the result of one toss of a fair die?

2. What is the expected value of the sum of two tosses of a fair die?

3. Guess the expected value of the sum of 10 tosses of a fair die.
    Does it seem reasonable to try to compute this value by direct use of the
    mathematical definition of expectation? A simple way to do this will be
    presented shortly.

---

1. The integral in Equation 4.28 is then the value that the integral

$$\int_{-n}^{n} xf_X(x)\,dx$$ (which is the area under the $xf_X(x)$ graph and over the interval from $-n$ to n) gets close to as n gets large.

4. In the state lottery suppose that a ticket costs $1.00, and that it is a winning ticket with a payoff of $1000 with a probability of .0005, otherwise it pays nothing back. What is the expected net gain? Remember to include the $1.00 paid for this ticket in the computation, and note that the net gain may be negative. (This is a fairly realistic assessment of how well state lotteries pay off — for the state.)

One question worth investigating is that of how close to the mathematical expectation you expect the average of n independent observations to be. To help provide a satisfactory answer to this question, we need to discuss numerical population variance, and its square root, the population standard deviation.

We introduced the concepts of sample standard deviation and population standard deviation for the case of a finite numerical population. We need to extend this concept to more general numerical populations. No matter what kind of numerical population being considered, we want to determine what the sample variance

$$s_n^2 = \frac{1}{n-1} \sum_{i=1}^{n} (X_i - \bar{X})^2 \qquad \text{4.29}$$

approaches for large sample size, n, where $\bar{X}$ is the sample mean of the n observations $X_1, ..., X_n$. But since the sample mean, $\bar{X}$, approaches the mathematical expectation, $E(X)$, for large sample size, n, and it does not matter much whether the division on the right of Equation 4.29 is by n or $n - 1$, it seems evident that for all populations where the sample variance approaches anything, it approaches the mathematical expectation

$$E([X - \mu]^2),$$

where $\mu = E(X)$.

With this reasoning in mind we give the following.

Definition 4.33 : Variance and standard deviation

The variance, $Var(X)$, of the random variable X (see Definition 4.26 on page 125) is defined by the equation

$$Var(X) = E([X - \mu]^2) \qquad \text{4.30}$$

where $\mu = E(X)$ (see Definition 4.30 on page 128, Definition 4.31 on page 129 and Definition 4.32 on page 130). The *standard deviation of X,* denoted by σ_X or just σ is the non-negative square root $\sqrt{var(X)}$.

❑

You might believe, from the definition of variance and expectation, that in order to compute the variance, you would first have to calculate the probability distribution of the random variable $Y = (X - \mu)^2$, and then compute $E(Y)$ as specified in the appropriate formula for variance. Fortunately, this is not the case. The following result, which simplifies the computation, can be established.

Theorem 4.34: Variance computations

Let μ stand for the expected value, $E(X)$, of the random variable X. For the case of X taking on only a finite number of values, $x_1,...,x_k$ with probabilities $p_1,...,p_k$ respectively, we find

$$var(X) = \sum_j (x_j - \mu)^2 p_j. \qquad 4.31$$

In the infinite discrete case, formula 4.31 also holds, provided the sum is finite.

The formula for the density case is

$$var(X) = \int_{-\infty}^{\infty} (x - \mu)^2 f_X(x)\, dx \qquad 4.32$$

provided this value is finite.

❑

Problem 4.35

Suppose g is a numerical valued function of a single numerical variable. If X is a random variable taking on only a finite number of values, $x_1,...,x_k$ with probabilities $p_1,...,p_k$ respectively, and Y is the random variable defined by $Y = g(X)$, show that

$$E(Y) = \sum_{i=1}^{k} g(x_i)p_i$$

Hint: Just group together those x_i giving rise to the same value of $g(x)$. This then justifies the finite valued X formula for $var(X)$.

Practice Exercises 4.28

1. Determine $E(X)$, $Var(X)$ and the standard deviation σ_X where X is the random variable giving the value on the toss of a fair die.

2. Same as previous question for a loaded die whose probabilities are respectively .1, .1, .2, .3, .15, .15.

3. Referring to the state lottery, see Practice Exercises 4.27.4. on page 131, what are the variance and standard deviation of the net payoff?

4. Referring to Example 4.26 on page 130, what are the variance and standard deviation of a Bernoulli random variable?

4.8 The Central Limit Theorem and its applications

The variance is useful in establishing theoretically, just how close you expect the sample mean and the (numerical) population mean to be.[1] The first result we present in this regard is a precursor to, and resembles the Central Limit Theorem, but is an inequality rather than an approximation.

Theorem 4.36: Chebychev inequality

Suppose X is a random variable whose expectation, $\mu = E(X)$ and variance, $\sigma^2 = Var(X)$ are well defined. Then for all $t > 1$,

$$P\{|X - \mu| \geq t\sigma\} \leq 1/t^2$$

$\square$

This states that for all positive numbers, t, the probability of the random variable, X, taking on a value whose distance from its expected value, μ, is at least t standard deviations, does not exceed $1/t^2$.

Suppose $\sigma = 3.2$.

If we choose $t = 4$, then the chances that X is further from μ than the distance $t\sigma = 12.8$ does not exceed $1/t^2 = 1/16$.

If we choose $t = 10$, we find that X has not more than a 1% chance of being further than 32 units from its expectation, μ.

We'll only establish this result for the simplest case — where X can take on only a finite number of values, with the following computation.

$$\sigma^2 = \sum_{\text{all } i} (x_i - \mu)^2 p_i \geq \sum_{\substack{i \text{ for which} \\ |x_i - \mu| \geq t\sigma}} (x_i - \mu)^2 p_i$$

$$\geq \sum_{\substack{i \text{ for which} \\ |x_i - \mu| \geq t\sigma}} (t\sigma)^2 p_i$$

$$= t^2\sigma^2 \sum_{\substack{i \text{ for which} \\ |x_i - \mu| \geq t\sigma}} p_i$$

$$= t^2\sigma^2 P\{|X - \mu| \geq t^2\sigma^2\}$$

---

1. The development here is a trifle circular, in that it uses probabilities to describe probabilities. Nonetheless, it has proved quite useful.

So $\sigma^2 \geq t^2\sigma^2 P\{|X - \mu| \geq t^2\sigma^2\}$, which, after dividing by $t^2\sigma^2$ proves the Chebychev inequality.

∟

The Chebychev inequality holds universally, no matter what the probability distribution of X, so long as it has a finite standard deviation. However it is an extremely conservative result because it has to hold for the worst case.

Here are a pair of results which simplify some of the most important expectation and variance computations.

Theorem 4.37: Expectation and variance of sums

Let $X_1, X_2,...,X_n$ be a sequence of independent random variables with finite expectations and variances, and let $a_1, a_2,...,a_n$ denote fixed real numbers. Then

$$E\left(\sum_{i=1}^{n} a_i X_i\right) = \sum_{i=1}^{n} E(a_i X_i) \text{ and } Var\left(\sum_{i=1}^{n} a_i X_i\right) = \sum_{i=1}^{n} a_i^2 Var(X_i)$$

☐

Although we won't supply a valid proof of this result in this text, you can empirically convince yourself of its validity by using Monte Carlo simulation (assuming that the random number generator produces statistically independent observations). A proof can be found in Blum-Rosenblatt [1].

Practice Exercises 4.29

1. Suppose $X_1, X_2,...,X_n$ is a sequence of independent random variables with common expectation μ and common standard deviation σ. Let $\bar{X}_n$ denote the sample mean

$$\bar{Y}_n = \frac{1}{n} \sum_{i=1}^{n} X_i$$

Using Theorem 4.37 above, find $E(\bar{X}_n), var(\bar{X}_n)$, and $\sigma_{\bar{X}_n}$ (the standard deviation of $\bar{X}_n$).

Due to its importance we have put the answer in the footnote below.[1] Look at it after doing this exercise.

2. Using the results of Exercise 1, how large should n be so that the standard deviation of the sample mean will be reduced by a factor —

---

1. $E(\bar{X}_n) = \mu, Var(\bar{X}_n) = \sigma^2/n, \sigma_{\bar{X}_n} = \sigma/\sqrt{n}$

    a. of 10 from that of the standard deviation of a single observation?

    b. of k from that of the standard deviation of a single observation? (These results show that obtaining high accuracy in estimating the population mean generally requires very large sample sizes; i.e., very accurate estimation is usually extremely expensive if the only tool for doing this is gathering lots of data. It may be better to look for better measurements.)

    c. Suppose X is a Bernoulli random variable representing the number of successes on a single trial of an experiment whose probability of success is an unknown value, p. Find a number, n, of independent trials of this experiment, that will guarantee with probability at least .95, that the proportion of successes is within .01 of the probability of success.

    Hints: Use the Chebychev inequality and the results of Exercise 4.29.1. on page 134. You also need to either find the largest possible value of $p(1-p)$ or an upper bound for this quantity.

We now come to the main result of this section, the Central Limit Theorem. It provides the theoretical justification for the (all too) common assumption that so many random variables have a distribution which is approximately normal (Gaussian). The Central Limit Theorem is considered by many to be the crowning achievement of probability theory. Although it may seem complicated, it actually provides a tremendous simplification for calculating many important probabilities — albeit at the cost of only providing an approximation to the probabilities being sought.

Roughly speaking the Central Limit Theorem asserts that the distribution function (see the definition of CDF on page 81) of the sum of results of sufficiently many repeated independent numerical valued experiments has approximately a Gaussian distribution function, with expectation and variance equal to those of the specified sum. We now state some more or less precise versions of this theorem.

Theorem 4.38: Central Limit Theorem

Let $X_1, ..., X_n$ be a sequence of independent, identically distributed, random variables (iids, see page 128) with finite population mean, μ and population standard deviation σ (see Definitions 4.30, 4.31, and 4.32, starting on page 128, and Definition 4.33 on page 131). Let $\bar{X}_n$ represent the sample mean, i.e.,

$$\bar{X}_n = \frac{1}{n} \sum_{i=1}^{n} X_i.$$

Then for sufficiently large sample size, n, the sample mean, $\bar{X}_n$ has approximately a Gaussian distribution function with mean μ and standard deviation $\sigma / \sqrt{n}$. That is to say, for each fixed constant c, and sufficiently large n (dependent on c and the distribution of the X_i)

$$P\{\overline{X}_n \le c\} \cong \Phi(c;\mu, \sigma/\sqrt{n})$$

4.33

where the right side of this approximation stands for the normal CDF (cumulative distribution function, see page 81) with mean μ and standard deviation $\sigma/(\sqrt{n})$ evaluated at the point c.

A more precise version is the following. For each fixed constant c, as n grows large

$$P\left\{\frac{\overline{X}_n - \mu}{\sigma/\sqrt{n}} \le c\right\} \quad \text{approaches} \quad \Phi(c)$$

4.34

where $\Phi(c)$ is the value of the standard normal CDF at c.

The classical version of the Central Limit Theorem is as follows. As n grows large, for each fixed real number, c,

$$P\left\{\frac{1}{\sqrt{n}} \sum_{i=1}^{n} \frac{X_i - \mu}{\sigma} \le c\right\} \quad \begin{array}{l} \text{approaches} \quad \Phi(c) \\ \text{(the value of the standard} \\ \text{normal CDF at } c). \end{array}$$

4.35

❑

Example 4.30: Wristwatch computations

Suppose your watch gains on the average 1.5 seconds per day, and deviates from this by no more than .2 seconds in any one day. We let T_i denote the actual amount of time your watch gains on day i (a gain may be negative). We ask for the probability that your watch gains more than 49 seconds in the month of September. We will translate the assumption about error to the equality $2\sigma = 0.2$. Then the total amount of time gained in September may be written as

$$\sum_{i=1}^{30} T_i.$$

We want to compute

$$P\left\{\sum_{i=1}^{30} T_i > 49\right\} = 1 - P\left\{\sum_{i=1}^{30} T_i \le 49\right\}$$

$$= 1 - P\{\overline{T}_n \le 49/30\}.$$

From the original assumptions above, we have $\mu = E(T_i) = 1.5$, and $\sigma = 0.1$. From Approximation 4.33 on page 136, we find the last written probability to be approximately

$$\Phi(49/30, 1.5, 0.1/\sqrt{30}) = \Phi(1.633, 1.5, 0.01826).$$

Putting the value of c = 1.633 in column c1, and then using the Minitab menu bar: Calc -> Probability distributions and choosing Cumulative probability, mean 1.5, standard deviation .01826, we obtain for the right hand side above, the value 1 (to 3 decimals). So, to 3 decimal places, there is less chance than 1 in 1000 that your watch will gain more than 49 seconds in September.

Similarly, if we change the 49 to 46, the approximation we get to the probability that your watch gains more than 46 seconds is .034.

■

Example 4.31 : Repeated Bernoulli trials

If you independently perform an experiment whose probability of success is .15, 100 times, approximately what is the probability of fewer than 11 successes?

Here we want to compute $P\left\{\sum_{i=1}^{100} X_i \le 10\right\}$, where $\mu = E(X_i) = 0.15$ and

$\sigma_{X_i} = \sqrt{0.15 \times 0.85} \cong 0.3571$. The event $\sum_{i=1}^{100} X_i \le 10$ can be rewritten as follows, so as to make use of the form of the Central Limit Theorem given by 4.35 on page 136.

$$\left\{\sum_{i=1}^{100} X_i \le 10\right\} = \left\{\frac{1}{\sqrt{n}} \sum_{i=1}^{100} \frac{X_i - \mu}{\sigma_{X_i}} \le \frac{10 - 100\mu}{\sqrt{n}\sigma_{X_i}}\right\}$$

$$= \left\{\frac{1}{\sqrt{n}} \sum_{i=1}^{100} \frac{X_i - \mu}{\sigma_{X_i}} \le \frac{10 - 100 \times 0.15}{10 \times 0.3571}\right\}$$

$$= \left\{\frac{1}{\sqrt{n}} \sum_{i=1}^{100} \frac{X_i - \mu}{\sigma_{X_i}} \le -1.400\right\}.$$

From the Central Limit Theorem (see expression 4.35 on page 136), the probability of this event is approximately the value of the standard normal distribution function at -1.4, $\Phi(-1.4) = 0.0808$; so the approximation given by the central limit theorem for the probability of less than or equal to 10 successes in 100 independent trials each of which has probability .15 of success on any one trial is about .0808. The exact probability of this event, computed from the binomial distribution with $n = 100$ and $p = .15$ is .0994. There are corrections which can improve the given approximation, but since the binomial distribution is now readily available for sample sizes up to 1000, where the Central Limit Theorem is really needed (for larger sample sizes than 1000), the Central Limit Theorem approximation is usually sufficiently accurate.

■

The first time someone meets the Central Limit Theorem, it seems quite surprising. Prior to the widespread availability of computers and statistical programs, the Central Limit Theorem was vital for reasonable approximations to the distributions of sums of independent, identically distributed random variables with a finite variance. It is still one of the jewels of probability theory, but is no longer critical for everyday computations. It remains as the theoretical cornerstone of the common assumption of normality of many measured variables.

It is interesting to investigate the basic reason that the Central Limit Theorem holds. We start by looking at the binomial distributions with $p = .5$. These are the distributions of the number of heads in n independent tosses of a fair coin. The probability of any particular sequence of n tosses with k heads is $(0.5)^k (0.5)^{n-k} = (0.5)^n$, independent of the value of k. The number of head-tail sequences of length n having k heads is the number of subsets of size k of the set of n available positions (used for the Heads). Thus, from Theorem 4.14 on page 108 **the probability of exactly k Heads in n independent fair tosses is** simply

$$\binom{n}{k}\left(\frac{1}{2}\right)^n . \qquad\qquad 4.36$$

Examining the value of expression 4.36 as k varies, for fixed large n, we see that it has the bell shape of the normal density. This happens because there is only one way in which $k = n$ heads can be achieved in n tosses — namely all heads; there are only n ways in which there can be exactly one head (one for each possible position of that head). But for $k = n/2$ there are a huge number of different head-tail sequences yielding k heads. It may still be surprising that the formula for the standardized normal density is

$$\frac{1}{\sqrt{2\pi}} e^{-x^2/2}$$

but the *bell shape* should be expected.

The same argument would seem to apply to sums of any number of independent, identically distributed random variables, because in a sense, the *size* of those sequences whose elements can add up to a *near central value* would be expected to decrease rapidly as this value moves away from the central value toward the extremes. Hence, the bell shape is not really unexpected in general, when adding up independent identically distributed random variables.

That is, it would appear that the Central Limit Theorem really depends on a *weight of numbers* argument.

Practice Exercises 4.32 Applying the Central Limit Theorem

1. A pair of fair dice is rolled 50 times.

    a. What are the largest and smallest possible sums?

    b. Approximately what is the probability of the sum being less than 380?

    c. Approximately what is the probability of the sum exceeding 340?

    d. Approximately what is the probability that the sum will lie between 335 and 376?

2. Suppose that a particle coming out of the smokestack undergoes random displacements in the north-south direction every minute, where these displacements have mean value .1 mile north, and standard deviation .1 mile.

    Approximately what is the probability that a particle will be further than 20 miles north of the smokestack,

    a. after 1 hour;

    b. after 2 hours;

    c. after 3 hours;

    d. after 3 hours and 10 minutes;

    e. after 5 hours.

    f. Where will most of the particles be located after 12 hours? (Furnish the shortest interval you can, which will include 99% of the particles.)

3. An experiment is assumed to have probability .2 of success and .8 of failure. The experiment is run 100 times. Approximately what are the chances of more than 15 successes? Compare this with the exact probability.

4. A research physician claims that his treatment has probability .9 of success. He tries the treatment on 400 people and observes 340 successes? Does his claim seem valid? Explain.

5. A manufacturing process has probability .2 of producing an acceptable unit. How many units should be manufactured so that you are approximately 95% certain of obtaining at least 200 good units?

Problems 4.39

1. One of the first problems that I had to solve after I started teaching came when the large midwestern school that employed me required all of the less distinguished faculty to attend graduation ceremonies. The date was early June, the heat was fierce, and full regalia (academic gown, mortar board, and black trousers) was required. Rented gowns were heavy and stifling, purchased ones were expensive, heavy, and stifling, and to compound the insult, the faculty involved had to bear all costume costs.

Rental was exorbitant, especially on the generous 5K/year I was paid. My wife, Lisa, came to the rescue with a flash of inspiration — *why not make the academic gown herself?* She purchased some very cool, black medium weight cotton material, and the pattern for a choir robe, and then started work. Everything went smoothly, until the time came to fit the pleated part of the sleeves and the back into the body of the robe. There were 136 pleats. Lisa sewed the pleats and tried to fit them to the robe. Too large. So she undid and redid the pleats. Too small. A few more tries and she requested assistance. My first probability problem. The first question was *how big were the pleats supposed to be?* The answer was *about 1/4 inch.* Next question, *how much error is there in the sewing of each pleat?* With the answer *at most 1/8 inch.* Then *can you tell if you made a pleat too big?* Answer *no.* Armed with this information, the assumption was made that the individual pleat lengths were uniformly distributed on the interval (1/8; 3/8) inches, and statistically independent. The desired total pleat length was 34 inches.

- a. What is the approximate distribution of the total pleat sewing error?
  Hints: The error in an individual pleat length is uniform on the interval (-1/8,1/8). You can estimate its standard deviation, σ, by Monte Carlo methods if you don't know calculus; $\sigma = \sqrt{2/192}$.

- b. Approximately what is the probability that the pleated part will fit into the other part of the robe, if you cannot tolerate a total error in pleat length of more than .3 inch?

- c. What conclusion can you draw if you know that any more than three tries will ruin the robe?
  We concluded that it would be best to do 1/5 of the pleats, then measure the partial length, and try to compensate on the next 1/5 of the pleats, etc. It worked.

2. Assume that your watch gains on the average of 1.5 seconds per day, with a standard deviation of $\sigma = 0.1$ seconds per day. Approximately how many days should elapse before resetting the watch, if we want to wait until the probability of a gain of at least 60 seconds has first exceeded .05?

3.. Suppose we have n independent Bernoulli trials, each with probability p of success and probability $1 - p$ of failure. The total number of successes in these n trials is a random variable having the binomial distribution with parameters n and p.

- a. Determine the formula for the probability of exactly k successes in the n trials. (That is, compute the binomial probability function.)
  Answer: If S_n denotes the total number of successes,
  $$P\{S_n = k\} = \binom{n}{k} p^k (1-p)^{n-k}$$

Hints for solution:
The probability of any particular sequence of n trials with k successes and n-k failures (say the 1st k trials being successes) is, due to the statistical independence, $p^k(1-p)^{n-k}$. This expression must be multiplied by the number of choices for the positions of the k successes.

b. Determine the formula for the cumulative distribution function of a binomial random variable with parameters n and p.

To conclude we present (without proof) a result which is certainly not difficult to believe if you accept the Central Limit Theorem (Theorem 4.38 on page 135)

Theorem 4.40: Distribution of independent normal sums

Suppose $X_1, X_2,...,X_n$ is a sequence of statistically independent normal random variables, with means $\mu_1, \mu_2,...,\mu_n$ and standard deviations $\sigma_1, \sigma_2,...,\sigma_n$. If $a_1, a_2,...,a_n$ is a sequence of fixed real numbers, then the weighted sum

$$\sum_{i=1}^{n} a_i X_i$$

has a normal distribution with mean and standard deviation

$$\sum_{i=1}^{n} a_i \mu_i \quad \text{and} \quad \sqrt{\sum_{i=1}^{n} a_i^2 \sigma_i^2} \quad \text{respectively}$$

(this mean and standard deviation can be calculated from Theorem 4.37 on page 134). ❑

Problems 4.41

1. If S denotes the sample space for some experiment and $\bigcup_{i=1}^{1000} A_i = S$ with $P\{B|A_i\} = 1/5$ for all i, what can you say about $P\{B\}$?

2. What are the advantages and disadvantages of the Central Limit Theorem versus exact computation of the behavior of a sum of random variables?

Review Exercises 4.33

1. Why is Monte Carlo simulation alone inadequate for a balanced understanding of the subject of probability?

2. Provide two different sample spaces for the toss of a pair of dice. Which one do you feel is preferable, and why?

3. What is meant by the phrase *the event A has occurred*?

4. On the toss of a pair of dice,
      let F_i be the event *first die comes up i,* for $i = 1,...,6$
      and
      let S_j be the event *second die comes up j,* for $j = 1,...,6$.
    In terms of these events write the events

    a. *first die comes up the same as the second die,*

    b. *first die comes up different from second die,*

    c. *first die result exceeds second die result?*

5. Referring to Exercise 4, compute the probabilities of all of the specified events using the usual uniform 36 point sample space model.

6. Why are proofs needed for proving results such as
    $P\{A \cup B\} = P\{A\} + P\{B\} - P\{A \cap B\}$?

7. What operations are used to create events from other events?

8. If all you knew was that the incidence of childhood leukemia was .00004, and that of polio was .00001, what can you legitimately say about the chances that a child would contract

    a. at least one of these diseases?

    b. both of these diseases?

9. In what part(s) of probability is counting important?

10. To determine the probability of *five spades, three hearts, four diamonds and one club* what is a convenient way to represent this event?

11. In statistically independent repetitions of an experiment with probability .9 of success, what is the probability of a success on the 1000th repetition, given that the previous trials were

    a. all successful?

    b. all failures?

    If you observed 999 successes on the first 999 trials of this experiment, what would you expect on trial 1000?

Chapter 5
Introduction to Statistical Estimation

5.1 Methods of Estimation

In trying to decide how we want an estimator to behave, it seems evident that we would like an estimator (function of the observations) to be *close*, in some sense, to the quantity we want to estimate. But here we run into a problem. An estimator that we have in mind may sometimes be close to the quantity we want to estimate, and at other times quite far away. If we are trying to estimate the *best* choice of stock to buy, the *high-tech* stocks would have been good choices prior to the year 2000, but not at all good at the beginning of the year 2001. Even an answer to the apparently simpler question of which of two estimators is a better one, is generally not possible — owing to the fact that one estimator may be better in some circumstances, but not in others.

You might think that there would be an acceptable answer to the question of which of two estimators was better in some statistical sense, such as one estimator being more likely to be close to the quantity being estimated than the other one. But a simple example shows that this criterion will not lead anywhere unless we require a little more — for instance, the estimator 3 is likely to be closer to the value being estimated than most other estimators, when the quantity being estimated is 3. But 3 hardly seems like a particularly desirable estimator in general.

There are several competing approaches to estimation. One approach, based on Bayes' theorem (see Equation 4.21 on page 118) does produce estimators which are optimum in some sense; but this approach relies on often unsupportable assumptions.

An approach which has proved pretty popular is to look for the best estimator in some reasonably restricted class of estimators — for instance, look for the estimator with the smallest standard deviation, within the class of *unbiased* estimators.[1] Here *best* usually is taken to mean *having the smallest standard deviation*. This frequently leads to reasonable estimators, but sometimes leads to unreasonable ones.[2]

---

1. Let μ be a value determined uniquely by the distribution function of the observed data. An *unbiased* estimate of μ is a statistic Y (function of the data only) satisfying the condition that no matter what the distribution function of the data, $E(Y) = \mu$. Then, for each of the possible distribution functions of the data, if $Y_1, Y_2,...,Y_n$ are independent identically distributed observations of Y, as n grows large, $\frac{1}{n} \sum_{i=1}^{n} Y_i$ gets close to μ.

2. As in the case where we want to estimate the probability, p^2 of success on two independent runs of an experiment whose probability of success is p.

Topic 5.1: Maximum likelihood estimators

The most widely used method of estimation is that of *maximum likelihood*. We first explain maximum likelihood in the discrete case. Suppose that the discrete probability function of the data $X_1, X_2, ..., X_n$ is completely determined by the parameter ϑ. In the case where the X_i are independent Bernoulli variables (1 with probability p, and 0 with probability $1 - p$) the parameter $\vartheta = p$. For a single Bernoulli variable, X, you can verify that the following formula is equivalent to the description just given.

$$P\{X = j\} = \vartheta^j(1 - \vartheta)^{1-j} \text{ for } j = 0, 1$$

In this case, from the assumed statistical independence of the X_i we find (see Definition 4.24 on page 121) the probability function

$$P\{(X_1 = x_1) \cap (X_2 = x_2) \cap ... \cap (X_n = x_n)\} = p_{X_1, X_2, ..., X_n}(x_1, x_2, ..., x_n; \vartheta)$$

$$= \vartheta^{x_1}(1 - \vartheta)^{1-x_1}\vartheta^{x_2}(1 - \vartheta)^{1-x_2} \cdots \vartheta^{x_n}(1 - \vartheta)^{1-x_n}$$

$$= \vartheta^W(1 - \vartheta)^{n-W} \text{ where } W = \sum_{j=1}^{n} x_j$$

Provided $x_1, x_2, ..., x_n$ is the actual observed value of the sequence of random variables $X_1, X_2, ..., X_n$, the *maximum likelihood estimate of ϑ* (in this case ϑ is p) is the value of ϑ maximizing the probability above.

In general, if ϑ is a (possibly vector) quantity whose value completely determines the distribution of the random variables to be observed, in the discrete case, the **maximum likelihood estimate of ϑ is the value which maximizes the probability of what has actually been observed**. So *maximum likelihood* looks like a fairly reasonable method — and it can be shown that maximum likelihood estimates tend to have desirable properties.

We now take a brief look at the case where the random variables $X_1, X_2, ..., X_n$ have a *joint density*. By this we mean roughly that there is a function, $f_{X_1, X_2, ..., X_n}$ such that the following approximation is a good one.

$$P\left(\bigcap_{i=1}^{n} \{x_i \le X_i \le x_i + h_i\}\right) \cong f_{X_1, X_2, ..., X_n}(x_1, x_2, ..., x_n)h_1 h_2 \cdots h_n \qquad 5.1$$

for each n-dimensional point $x_1, x_2, ..., x_n$, provided all of the h_i have sufficiently small magnitude. The function $f_{X_1, X_2, ..., X_n}$ (often abbreviated as simply f, i.e., omitting the identifying indices) is called the *joint probability density of the random variables* $X_1, X_2, ..., X_n$. In the statistical case, we usually deal with a class of probability density functions, in which case, for each value of the parameter ϑ we would have a distinct joint probability density whose value at ϑ is denoted by $f_{X_1, X_2,...,X_n}(x_1, x_2,...,x_n;\vartheta)$. The parameter ϑ might be a vector, such as (μ, σ) in the case of independent normally distributed random variables.

In this case **the maximum likelihood estimate of** ϑ is the value of ϑ maximizing $f_{X_1, X_2,...,X_n}(x_1, x_2,...,x_n;\vartheta)$, where $x_1, x_2,...,x_n$ is the actual observed value of the sequence of random variables $X_1, X_2,...,X_n$.

In most cases the actual process of finding maximum likelihood estimates requires mathematical methods beyond the level of this text (as does the process of finding *unbiased estimates having smallest standard deviation*, mentioned a bit earlier).

These mathematical limitations make it unfeasible to investigate the method of maximum likelihood here, although they are absolutely necessary for developing models to cope with situations that have not been previously encountered. However, they will not prevent us from becoming familiar with the most commonly used estimators employed by scientists who need statistics for analyzing their data.

Topic 5.2: Natural estimators

One class of what we might call *natural estimators* arises if the unknown parameters (which determine the possible probability distributions of the data) are known, simple functions of the **moments**, $m_k = E(X^k)$ for $k = 1,2,...,K$ (where K is known and not too large). A case we have already met is the variance — i.e.,

$$\vartheta = var(X) = E[(X-\mu_1)^2] = E(X^2) - [E(X)]^2 \qquad 5.2$$

To estimate the variance from a sample of independent observations on X, we could use the natural estimator

$$\frac{1}{n}\sum_{i=1}^{n} X_i^2 - \left[\frac{1}{n}\sum_{i=1}^{n} X_i\right]^2$$

More generally if a component of the vector parameter ϑ has the form $g(\mu_1, \mu_2,...,\mu_K)$, where g is a well-behaved function, then a natural estimator for this component is $g(m_1, m_2,...,m_K)$ where for $k = 1,2,...,K$ the k-th *sample moment* m_k is given by

$$m_k = \frac{1}{n}\sum_{i=1}^{n} X_i^k \qquad 5.3$$

The approach to estimation just given is called *the method of moments*.

Another natural estimator arises if we want to estimate a component of the (possibly vector) parameter ϑ, and this component is a function of some finite sequence of quantiles — one example being the interquartile range $q_{.75} - q_{.25}$. Here we might use the difference $q_{.75} - q_{.25}$ of the 75-th and 25-th *sample percentiles* as our estimate of the interquartile range.

Just as with the method of moments presented on the previous page, suppose that the quantity of interest that we want to estimate is a known well-behaved function $Q(q_{\pi 1}, q_{\pi 2}, ..., q_{\pi K})$, where $q_{\pi 1}, q_{\pi 2}, ..., q_{\pi K}$ are specified **quantiles**[1] or **percentiles**. Based on our claims concerning sample medians on page 61, and made somewhat convincing by applying the macro of Problems 3.12.6. on page 66, it should be no surprise that the statistic $Q(q_{\pi 1}, q_{\pi 2}, ..., q_{\pi K})$ (where $q_{\pi i}$ is the sample percentile corresponding the population percentile $q_{\pi i}$) is frequently used to estimate this quantity.

5.2 Distribution of Sample Percentiles

To provide a firmer theoretical basis for such estimation, in the following paragraphs, we will show that the sample percentile tends to be close to the associated population percentile when the sample size is large enough.

It's easiest to precede the treatment of sample percentiles with a short discussion of order statistics.

Definition 5.3: Order statistics

Suppose we are given a sequence $X_1, X_2, ..., X_n$ of random variables. The *order statistics* associated with this sequence is a sequence denoted by $X_{[1]}, X_{[2]}, ..., X_{[n]}$. The value of $X_{[i]}$ is the value of the element of $X_1, X_2, ..., X_n$ which is *i*-th in increasing algebraic order. So

$$X_{[1]} \le X_{[2]} \le \cdots \le X_{[n]}.$$

It should be noted that repeated values in the $X_1, X_2, ..., X_n$ sequence yield repeated values in the order statistics sequence $X_{[1]}, X_{[2]}, ..., X_{[n]}$.

❏

The derivation of the distribution of sample percentiles is not particularly difficult or even tedious, provided we are a bit cavalier about certain subtle points — namely the definition of the sample 100π-percentile.

---

1. Recall the median, which is the 50-th percentile, or .5 quantile, which was discussed on page 60. In general, when π is a number between 0 and 1, the π-quantile (also called the 100π-percentile), q_π of a population is essentially the value dividing the population so that 100π % of the population is to the left of q_π and $100(1-\pi)$% is to the right of q_π, with appropriate modifications if q_π has nonzero probability. The sample π-quantile (also called the 100π sample percentile), q_π divides the (ordered) sample in the same way.

Definition 5.4: Sample percentiles (quantiles)

For π some given probability with $0 < \pi < 1$, the *sample* 100π *percentile* (also called the *sample* π-quantile) is simply the order statistic $X_{[n\pi]}$ where $\underline{n\pi}$ is the closest integer to $n\pi$ which does not exceed $n\pi$. This definition may differ from other definitions of the sample π-quantile, but in inconsequential ways.

$\square$

From this definition, we see that the key to determining properties of the percentiles lies in determining the distribution of the order statistics, $X_{[1]}, X_{[2]},...,X_{[n]}$, related to the sequence $X_1, X_2,...,X_n$. To get a handle on this problem, we notice that the sample median is less than a number q if and only if at least half of the observations are less than q. This idea can be extended by defining the Bernoulli random variables Y_i, $i = 1, 2,..,n$ as follows.

$$Y_i = 1 \ \text{ if } \ X_i \le q \ \text{ and } \ Y_i = 0 \ \text{ if } \ X_i > q \ . \qquad\qquad 5.4$$

Then you can see that for each integer j and each real number q

$$X_{[j]} \le q \quad \text{if and only if} \quad S_n \ = \ \sum_{i=1}^{n} Y_i \ge j.$$

But because $X_i \le q$ if and only if $Y_i = 1$, and $P\{X_i \le q\} = F(q)$ (where F is the cumulative distribution function of the Y_i (see page 81)),

$$S_n \ = \ \sum_{i=1}^{n} Y_i$$

has the binomial distribution (see page 30) with parameters n and $p = F(q)$.

Hence, for the j-th order statistic of interest, we find

$$P\{X_{[j]} \le q\} \ = \ P\{S_n \ge j\} \ = \ 1 - P\{S_n \le j - 1\} \qquad\qquad 5.5$$

where S_n has the binomial distribution with parameters n and $p = F(q) = P\{X_i \le q\}$. For reference we state this result formally.

Theorem 5.5: Distribution of order statistics

Let $X_{[j]}$ be the j-th order statistic (Definition 5.3 on page 146) from the sequence $X_1, X_2,...,X_n$ of independent random variables all having cumulative distribution function F (page 81). Then

$$P\{X_{[j]} \le q\} \ = \ P\{S_n \ge j\} \ = \ 1 - P\{S_n \le j - 1\}$$

where S_n has the binomial distribution with parameters n and $p = F(q) = P\{X_i \le q\}$.

$\square$

Applying this result, and the Central Limit Theorem (Theorem 4.38 on page 135) to the sample 100π-percentile, $X_{[n\pi]}$ (Definition 5.4 on page 147) and the population 100π-percentile, q_p, we have the approximation

$$P\{X_{[\underline{n\pi}]} \le q\} = 1 - \{S_n \le \underline{n\pi} - 1\}$$

$$= 1 - P\left\{\frac{S_n - np}{\sqrt{np(1-p)}} \le \frac{n\pi - 1 - np}{\sqrt{np(1-p)}}\right\}$$

$$= 1 - P\left\{\frac{S_n - nF(q)}{\sqrt{nF(q)(1-F(q))}} \le \frac{n\pi - 1 - nF(q)}{\sqrt{nF(q)(1-F(q))}}\right\} \qquad 5.6$$

$$\cong 1 - P\left\{Z \le \frac{n\pi - 1 - nF(q)}{\sqrt{nF(q)(1-F(q))}}\right\}$$

where Z is a standard normal random variable. From this approximation, it is not difficult to see that if $F(q) < \pi$ then $P\{X_{[n\pi]} \le q\}$ gets close to 0 for large n, while if $F(q) > \pi$ then $P\{X_{[n\pi]} \le q\}$ approaches 1 for large n. Thus if we choose q_L with $F(q_L) < \pi$ and q_U with $F(q_U) > \pi$, we see that for large n, $X_{[n\pi]}$ will have a high probability of being in the interval (q_L, q_U). This shows that for large enough n, the sample π-quantile tends to be close to the population π-quantile, as asserted. We can get an exact idea of the probability that the sample π-quantile lies in some interval by using the binomial distribution for this determination. Using Minitab, in some chosen column put the values of the qs you want to use in the binomial computation in the first line of computations 5.6 above (values of $\underline{n\pi} - 1$). Then on the menu bar choose

    Calc -> Probability Distributions -> Binomial -> Cumulative probability. Fill in the column number (e.g., c1) you used for the qs .

Practice Exercises 5.1

1. Let $X_1,...,X_{50}$ be a sample from any distribution with unique quantiles, q_p for all p with $0 < p < 1$. Find $P\{X_{[50 \times 0.9]} \le q_p\}$ for $p = .95, .975$.

2. What does the result of Exercise 1. tell you about how you can use such a sample to learn something about the unique quantiles of any distribution from which this data arose?

Problems 5.6

1. Compute the probability that the sample median from a standard normal distribution based on 100 observations is within .15 units of the true median, and compare this probability with the probability that the sample mean from a standard normal distribution based on 100 observations is within .15 units of the true mean. (Both parameters equal 0.) What does this seem to tell you about which is the better estimator?

2. Let $\bar{X}_{225}$ be the sample mean of 225 observations from the standard normal distribution. Compute the probability that $\bar{X}_{225} + 1$ is within .1 units of the 84.13-th percentile of the standard normal distribution. Compute the probability that the sample 84.13-th percentile of the standard normal distribution is within .1 unit of this distribution's 84.13-th percentile. What, if anything, was learned from these computations.

3. For the Poisson distribution with mean λ, the variance is also λ. First, using the random number generator, see if this seems to be true, based on samples of size 100, and $\lambda = 1, 3, 10$. Which of the two natural estimates (see equation 5.2 on page 145) of λ seems better? Is there convincing evidence of the validity of your answer? (Soon we'll see how to obtain more convincing evidence.)

5.3 Adequacy of Estimators

Usually, in situations involving estimation, we have some readily available choices of estimates, and want to determine first if any of them are adequate for the problem or problems we want to solve. By *adequate* we will usually mean whether or not the estimator we choose is likely to be *close enough* to the quantity being estimated, to do the job we have in mind. One approach would be to calculate the probability distributions of the estimates of interest and try to calculate whether any of them is likely to be close enough to the quantity being estimated. This approach is likely to take more effort than we can afford, even if we are capable of carrying it out. But there are other alternatives, which may still be acceptable, and may be much easier to implement. One of these is suggested by the Central Limit Theorem (Theorem 4.38 on page 135) because it is often the case that (with this theorem as justification) the estimates we are considering are either approximately normally distributed, or some simple function of them is approximately normal. This happens quite a bit more often than you might think. Even if the assumption of normality is not acceptable, the Chebychev inequality (Theorem 4.36 on page 133) might apply, and may yield a usable result.

If the Central Limit Theorem applies, and the estimator you are examining is therefore approximately normally distributed, you may be exceptionally lucky and know its standard deviation, denoted by σ. This could be the case if you are using a measuring instrument that you are very familiar with. Then, if you want to be reasonably sure that your estimate, denoted by Y, is *close enough*, say, within k units of its expected value, μ, that you want to estimate, you simply compute the probability, $P\{|Y - \mu| < k\}$ and see whether it is *large enough*. This can be done using the Minitab menu bar, choosing Calc -> Probability distributions, etc., as shown in Example 2.2 on page 26.

You may not be quite so lucky, even if the Central Limit Theorem applies, but you might be able to come up with a **worst case**, i.e., you might know that the standard deviation does not exceed some known value, σ_0. By acting as if the actual standard deviation was really σ_0, you can still determine whether your estimate is good enough to handle all possible cases. Here are some illustrative examples.

Example 5.2: Is your blood pressure monitor good enough?

Suppose you have a blood pressure monitor which gives you readings that are unbiased (i.e., have the desired expected value) but have a standard deviation of 15 points. For various reasons you can only make 9 independent readings to try to determine your blood pressure. You would like to be 90% certain that the average of these 9 readings is within 10 points of your expected blood pressure. Is your monitor adequate for this purpose? Assuming either that the individual readings are independent each with a normal distribution, or that the Central Limit Theorem applies to this device, leads to the conclusion (see Theorem 4.37 on page 134 and Theorem 4.40 on page 141) that the average

$$\bar{X}_9 = \frac{1}{9} \sum_{i=1}^{9} X_i$$

has (at least approximately) a normal distribution with your expected blood pressure, μ, as its mean and standard deviation of 5 points. To answer the given question, we want to compute

$$P\{|\bar{X}_9 - \mu| \le 10\} = P\{-10 \le \bar{X}_9 - \mu \le 10\}.$$

Because $E(\bar{X}_9 - \mu) = 0$ and $Var(\bar{X}_9 - \mu) = Var(\bar{X}_9) = \frac{15^2}{9} = \left(\frac{15}{3}\right)^2 = 5^2$

all we need to determine is the probability that a normal random variable with mean 0 and standard deviation 5 takes on a value between -10 and 10. Using the method of Example 2.2 on page 26, we find this probability to be .9544. That is, you are more than 95% sure that the average of 9 readings is within 10 points of the *true* blood pressure. Hence we are more than 90% sure of this event — and, for our purposes, the given blood pressure device is good enough.   ∎

Example 5.3: Is a pure random sample of 300 adequate?

In a presidential election poll you want to see if you can be 95% sure that the proportion of people in a pure random sample of 300 for the Democratic candidate is within .03 of the population proportion of people for the Democrat. To handle this question, we note that the proportion of democratic votes in the sample is

$$\frac{1}{300} \sum_{i=1}^{300} X_i$$

where X_i is the number of democratic votes from person i in the sample (0 or 1).

We want to compute the probability

$$P\left\{ \left| \frac{1}{300} \sum_{i=1}^{300} X_i - p \right| \le 0.03 \right\}$$

The Central Limit Theorem (Theorem 4.38 on page 135) allows us to obtain a good approximation to this probability for any choice of p. To put this expression in Central Limit Theorem form, rewrite it as

$$P\left\{\left|\frac{1}{\sqrt{300}}\sum_{i=1}^{300}\frac{X_i-p}{\sqrt{p(1-p)}}\right|\le\frac{0.03\times\sqrt{300}}{\sqrt{p(1-p)}}\right\}$$ 5.7

By the Central Limit Theorem, the random variable within the absolute value signs has a distribution function which is approximately standard normal. Acting as if this approximation is exact, it follows that the probability of Expression 5.7 is smallest when the denominator of the right side of the inequality is largest. So we need to determine the largest value of $p(1-p)$ for $0\le p\le 1$. Since $p(1-p)$ is a variance, we would expect this to occur at the value of p yielding the most uncertainty — namely, $p=.5$. This is the case, as we see from the following *completion of the square* computations.

$$p(1-p) = -(p^2-p) = -\left(p^2-p+\frac{1}{4}\right)+\frac{1}{4} = -\left(p-\frac{1}{2}\right)^2+\frac{1}{4}.$$

which is evidently largest when $p=1/2$, as guessed. Thus, at least approximately

$$P\left\{\left|\frac{1}{\sqrt{300}}\sum_{i=1}^{300}\frac{X_i-p}{\sqrt{p(1-p)}}\right|\le\frac{0.03\times\sqrt{300}}{\sqrt{p(1-p)}}\right\}\ge P\left\{\left|\frac{1}{\sqrt{300}}\sum_{i=1}^{300}\frac{X_i-p}{\sqrt{p(1-p)}}\right|\le\frac{0.03\times\sqrt{300}}{0.5}\right\}$$

$$\cong P\{|Z|\le 1.0392\}$$
$$= 0.7012$$

So we find that there is only about a 70% chance that the proportion of democratic voters in a random sample of 300 voters will be within .03 of the proportion of democratic voters in the population when this latter proportion is near 50%. Thus, when the public opinion pollers claim to be using a sample of size 300 to obtain such results, they must be doing something fancier than using a simple random sample from the voting population.

What are they doing? In essence, they are taking simple random samples from disjoint rather homogeneous subpopulations. In each of these subpopulations, the proportion of democratic voters is far away from 1/2. The subpopulations are chosen with a known history of such voting patterns in mind — e.g., African Americans and Jews voting democratic, fundamentalists voting republican, etc. This allows the *worst case* estimates of the standard deviation $\sqrt{p(1-p)}$ to be chosen much smaller than .5 for each of the subpopulations. The stratified sampling theorem, Theorem 4.20 on page 116, can then be applied, provided the subpopulation sizes are known. This method only works when disjoint subpopulations of known sizes can be found, and only works well when the proportion of interest is far from .5 in a sufficient proportion of these subpopulations.

A similar approach to improving results is seen when we look for the effect of one quantity (e.g., calories) on another (e.g., weight). Since other factors also influence weight, if the most important of these can be accounted for in the model, and measured, their use can often lead to a great improvement in results, without requiring great increases in sample size.

■

Practice Exercises 5.4

1. Compute the exact probability approximated by the result of Example 5.3.

2. Determine the exact sample size (using the binomial distribution) required for the sample proportion to be within .03 of a probability of .5 with probability .95.

3. Referring to Example 5.2 on page 150 how many observations would be needed if you wanted to be 99% sure that the average of 9 readings was within 5 points of the expected blood pressure?

5.4 Confidence Limits and Confidence Intervals

A **confidence interval** for a numerical parameter whose value is not known, is an interval constructed from the data which, for each possible parameter value, has at least some prescribed probability of including this parameter. An **upper [lower] confidence limit** for a numerical parameter whose value is unknown, is an observable random variable whose probability of being greater than or equal to [less than or equal to] the parameter is at least some specified probability. A confidence interval is thus determined by a pair of confidence limits.

More precisely we have:

Definition 5.7: 1- α level confidence limits and intervals

Suppose we can observe the values of random variables, $X_1, X_2,...,X_n$, and that their joint distribution function, $F_{X_1, X_2,...,X_n}$, specified completely by a (possibly vector) parameter ϑ, is given by

$$F_{X_1, X_2,...,X_n}(x_1, x_2,...,x_n;\vartheta) = P_\vartheta\left\{\bigcap_{i=1}^{n} X_i \leq x_i\right\}$$

(Here P_ϑ is a unique probability measure determined by the usually unknown parameter ϑ.) Random variables

$$Y_L = g_L(X_1, X_2,...,X_n) \quad \text{and} \quad Y_U = g_U(X_1, X_2,...,X_n),$$

not dependent on ϑ, are said to be $1 - \gamma$ **lower** and $1 - \delta$ **upper confidence limits** respectively, for the scalar value $h(\vartheta)$, if for all values of ϑ

$$P_\vartheta\{Y_L \leq h(\vartheta)\} \geq 1 - \gamma \qquad\qquad 5.8$$

and

$$P_\vartheta\{h(\vartheta) \le Y_U\} \ge 1 - \delta .$$ 5.9

Because the event $\{Y_L \le h(\vartheta) \le Y_U\}$ stands for the intersection

$$\{Y_L \le h(\vartheta)\} \cap \{h(\vartheta) \le Y_U\}$$

it follows from Bonferroni's inequality (Theorem 4.8 on page 103) that

$$P\{Y_L \le h(\vartheta) \le Y_U\} \ge 1 - \gamma - \delta$$ 5.10

and hence, the interval $[Y_L, Y_U]$ is called a $1 - \gamma - \delta$ level confidence interval for $h(\vartheta)$.

If the equalities hold respectively for the probabilities in Equations 5.8, 5.9, and 5.10, we refer to these as *exact confidence limits and an exact confidence interval.*

❑

There is some disagreement about how the value of any particular confidence interval should be interpreted. Suppose we are dealing with a confidence interval for which $\gamma + \delta = 0.05$, and that the observed values of $Y_L = 1.1$ and $Y_U = 3.4$. Some statisticians claim that once the values of Y_L and Y_U are observed, there is no longer a probability interpretation, since either $h(\vartheta)$ does lie in the interval given by these values, or it does not. In the case given, it is certainly true that either $h(\vartheta)$ lies between 1.1 and 3.4 or it fails to do so. The problem being overlooked here is that the central issue is the status of our information. If **every time** we compute the value off this confidence interval's end points, we make the assertion that the interval that arose included the value of $h(\vartheta)$, about 95% or more of our assertions would be correct.[1] So, lacking any other information about $h(\vartheta)$, we have every right to have good confidence that any instance of such an interval includes $h(\vartheta)$. If, on the other hand, we had created the data and chosen $h(\vartheta)$ to equal 3, we would know which of the instances of any of these intervals included $h(\vartheta)$ and which did not. But this is an example in which we do not lack initial information about $h(\vartheta)$, and it is in such situations that we cannot interpret the observed confidence interval in the usual way. In a nutshell, if we haven't stacked the deck it seems reasonable to have good confidence that a single observed confidence interval includes the *true value, $h(\vartheta)$.* Now we examine a slightly less clear situation. Suppose we repeated our experiment many times, and constructed exact 95% level confidence intervals for $h(\vartheta)$, and chose to observe whether or not the number 3 was in the intervals which were generated. If these intervals included 3 about 95% of the time, I would tend to believe that 3 was pretty close to $h(\vartheta)$, while if 3 was in about 85% of the intervals it would certainly seem to me that $h(\vartheta)$ was not equal to 3. With access to a whole sequence of repeated experiments, the fact that a particular interval included the value 3 cannot be used by itself to draw conclusions. But here again, we have extra information — namely that we are observing a sequence of identical (as far as we can control) experiments.

---

1. Assuming you subscribe to the frequency interpretation of probability.

Confidence intervals furnish a way to present a reasonably complete summary of the information that can be extracted from data about unknown statistical parameters. A typical situation is the following. Two treatments are being tested for use against some disease. Both are known to work, but if one of them acts faster than the other, it is preferred. To compare the two treatments, we get some data, and construct a .95 level *exact* confidence interval for the difference, $m_2 - m_1$, of their mean recovery times. Conventionally, if the confidence interval lies completely to the right of 0, it is concluded that treatment 1 is preferred to treatment 2; if the confidence interval lies completely to the left of 0, treatment 2 is preferred to treatment 1; if the confidence interval includes 0, neither treatment is chosen as preferred. But this conventional treatment is far too simplistic. There are about 4 basic situations which can be well conveyed by confidence intervals, and they are illustrated conceptually in Figure 5-1 below.

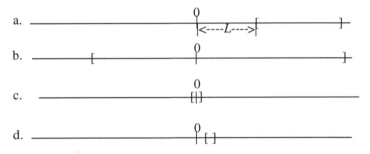

Figure 5-1   Situations Distinguished by Confidence Intervals

In case a. we see that the difference, $m_2 - m_1$, is statistically different from 0, and clinically different from 0 (assuming L in Figure 5-1.a represents a difference of clinical significance).

In case b., even though the difference, $m_2 - m_1$, is not statistically different from 0, there are values inside the confidence interval which would be significant clinically, i.e., clinically different from 0, if they were achieved. This is a case where we need more data to spot most differences that would be clinically significant.

In case c., it is not just that we fail to have any statistical evidence that the difference $m_2 - m_1$ is statistically different from 0. From a practical view, no value in the confidence interval is clinically different from 0. So in this type of situation we could say that from a practical viewpoint, $m_2 - m_1 = 0$.

Case d. represents the situation in which there is so much data that we spot a statistical difference from 0 which has no clinical significance. Here it would also be legitimate to claim that $m_2 - m_1 = 0$ from a clinical viewpoint.

Figure 5-1 is just a symbolic illustration, because the user must interpret the confidence interval values using his/her own judgement about their importance.

Topic 5.8: Elementary confidence interval construction

In the next few examples construction of some of the most commonly used confidence intervals is illustrated. First we need the following definition.

Definition 5.9: Standard normal distribution upper $1\text{-}\beta$ point

For β any number strictly between 0 and 1, $z_{1-\beta}$ is the unique value for which

$$\Phi(z_{1-\beta}) = 1-\beta,$$

where Φ is the standard normal cumulative distribution function (see page 81). The value $z_{1-\beta}$ is called the *upper $1-\beta$ point of the standard normal distribution*, or the $1-\beta$ *quantile of the standard normal distribution*, or *the* $100\,(1-\beta)$ *percentile of the standard normal distribution*. **This definition extends to any one-dimensional distribution with CDF** F (see page 81) **for which there is a unique value** $z_{1-\beta}$ **(the upper $1-\beta$ point of** F**) with** $F(z_{1-\beta}) = 1-\beta$.

$\square$

Example 5.5: Interval for normal mean with known standard deviation

We start with the assumption that we have n independent observations $X_1,...,X_n$ on some quantity (say red blood count) which are normally distributed with **unknown expectation** μ and **known standard deviation** σ. The known variability is assumed to arise from extraneous factors (such as temperature variation and variation in the amount of blood scanned by the measuring device) assumed under fairly tight control. The expectation is unknown because we anticipate different expectations from different patients, and the n observations are assumed to be from a single patient.

From Theorem 4.40 on page 141 it is known that the unobservable random variable $\sqrt{n}(X_n-\mu)/\sigma$ (where X_n is the sample mean of $X_1,...,X_n$) has a standard normal distribution. From this it follows, using Definition 5.9, that

$$P\{-z_{1-\alpha/2} \le \sqrt{n}(\bar{X}_n-\mu)/\sigma \le z_{1-\alpha/2}\} = 1-\alpha/2.$$

Using simple algebraic manipulation, the event

$$-z_{1-\alpha/2} \le \sqrt{n}(\bar{X}_n-\mu)/\sigma \le z_{1-\alpha/2}$$

can be rewritten as

$$\bar{X}_n - z_{1-\alpha/2}\sigma/\sqrt{n} \le \mu \le \bar{X}_n + z_{1-\alpha/2}\sigma/\sqrt{n}$$

(these manipulations being the following for the first inequality.

$$\sqrt{n}(\bar{X}_n-\mu)/\sigma \le z_{1-\alpha/2} \Leftrightarrow \bar{X}_n-\mu \le z_{1-\alpha/2}\sigma/\sqrt{n}$$

$$\Leftrightarrow \bar{X}_n \le \mu + z_{1-\alpha/2}\sigma/\sqrt{n}$$

$$\Leftrightarrow \bar{X}_n - z_{1-\alpha/2}\sigma/\sqrt{n} \le \mu$$

where the symbol $\Leftrightarrow$ stands for the phrase *if and only if.*

So,

$$P\{\sqrt{n}(\overline{X}_n-\mu)/\sigma \le z_{1-\alpha/2}\} = P\{\overline{X}_n - z_{1-\alpha/2}\sigma/\sqrt{n} \le \mu\}.$$

Thus we see that a $1 - \alpha/2$ level exact lower confidence limit for the mean, μ, of independent normal random variables $X_1,...,X_n$ with mean μ and known standard deviation σ is $\overline{X}_n - z_{1-\alpha/2}\sigma/\sqrt{n} \le \mu$.

∎

Topic 5.10: Summary of normal mean confidence results when standard deviation is known

If $X_1,...,X_n$ are independent normal random variables with unknown mean μ and known standard deviation σ then (keeping in mind the definition of confidence intervals and confidence limits, Definition 5.7 on page 152)

a $1 - \alpha/2$ level exact lower confidence limit for μ is $\overline{X}_n - z_{1-\alpha/2}\sigma/\sqrt{n}$

a $1 - \alpha/2$ level exact upper confidence limit for μ is $\overline{X}_n + z_{1-\alpha/2}\sigma/\sqrt{n}$

a $1 - \alpha$ level exact confidence interval for μ is $[\overline{X}_n-z_{1-\alpha/2}\sigma/\sqrt{n}\,;\overline{X}_n + z_{1-\alpha/2}\sigma/\sqrt{n}]$

where $z_{1-\alpha/2}$ is the upper $1 - \alpha/2$ point on the standard normal distribution (see Definition 5.9 on page 155).

❑

Before proceeding further, we take the time to examine exactly what was done in order to obtain the confidence limits and interval above. **We started off with an unobservable random variable**, $R(\mu)$ (here $R(\mu) = \sqrt{n}(\overline{X}_n-\mu)/\sigma$) whose probability distribution is completely known; $R(\mu)$ is unobservable because it depends on μ, σ being assumed known. Because the distribution of $R(\mu)$ is known, we can choose numbers, $a_{\alpha/2}$ and $a_{1-\alpha/2}$ such that

$$P\{a_{\alpha/2} \le R(\mu) \le a_{1-\alpha/2}\} = 1-\alpha.$$

If $R(\mu)$ always increases [or always decreases] as μ increases, we can solve the inequalities $a_{\alpha/2} \le R(\mu) \le a_{1-\alpha/2}$ to rewrite this event

(in the case of $R(\mu)$ increasing) $R^{-1}(a_{\alpha/2}) \le \mu \le R^{-1}(1-\alpha/2)$

(in the case of $R(\mu)$ decreasing) $R^{-1}(1-a_{\alpha/2}) \le \mu \le R^{-1}(\alpha/2)$

one of the above giving us the confidence interval we were seeking.

Problems 5.11

Construct confidence intervals for the indicated parameters from the given information.

1. If $X_1,...,X_n$ are independent normal random variables with unknown mean μ and unknown standard deviation σ, examine the unobservable random variable

$$t_n = \sqrt{n}(\bar{X}_n - \mu)/s_n,$$

where $\bar{X}_n$ is the sample mean and s_n is the sample standard deviation of $X_1,...,X_n$, given by

$$\bar{X}_n = \frac{1}{n}\sum_{i=1}^{n} X_i, \quad s_n = \sqrt{\frac{1}{n-1}\sum_{i=1}^{n}(X_i - \bar{X}_n)^2}$$

The variable t_n has the *student t distribution, with n - 1 degrees of freedom.* Find a confidence interval for the population mean, μ, based on $X_1,...,X_n$.

Hints: It may help to have symbols for the CDF (see page 81) of the t distributions, and for their percentiles.

2. Suppose $X_1,...,X_n$, $Y_1,...,Y_m$ are independent sequences of independent normal variables, with the X_i having unknown expectation μ_x and unknown standard deviation σ_x and the Y_j having unknown expectation μ_y and unknown standard deviation σ_y. Then the unobservable random variable

$$F_{n-1, m-1} = \frac{s_{x,n}^2/\sigma_x^2}{s_{y,m}^2/\sigma_y^2}$$

(where $s_{x,n}$ and $s_{y,m}$ are respectively the sample standard deviations of the sequences $X_1,...,X_n$ and $Y_1,...,Y_m$ [see Exercise 1. above]) has the (Fisher) F distribution with n-1 and m-1 degrees of freedom. Use this information to compute confidence limits and confidence intervals for σ_x^2/σ_y^2 and for σ_x/σ_y. What would you guess these confidence intervals get used for?

3. Suppose $X_1,...,X_n$ are independent normal variables, with the X_i having unknown expectation μ and unknown standard deviation σ. Let s_n be the sample standard deviation of the $X_1,...,X_n$ (see Exercise 1. above). Use the fact that the distribution of the unobservable random variable

$$(n-1)s_n^2/\sigma^2$$

has the Chi-square distribution with n - 1 degrees of freedom) to compute confidence limits and confidence intervals for σ^2 and σ.

4. Generate appropriate sequences of normal random variables to see what confidence intervals you get from the results of your answers to the three previous problems.

We should mention concerning the confidence interval for the mean, μ, of a normal random variable with unknown standard deviation (Problem 5.11.1. on page 156) that the price of not knowing the standard deviation is that no matter what the sample size, its length cannot be known prior to knowing the data. This means that the confidence interval may be too long to provide much useful information.[1]

Practice Exercises 5.6

1. Use the $1 - \alpha$ confidence interval $\bar{X}_n \pm t_{n-1,1-\alpha} s_n / \sqrt{n}$, for the mean of n identically distributed normal random variables, (gotten from the result of Problem 5.11.1. on page 156) where $t_{n-1,1-\alpha}$ is the upper $1 - \alpha$ point ($100 \times (1 - \alpha)$ percentile) of the t distribution with n -1 degrees of freedom, to generate $n = 15$ normal random samples of size 100 each, with mean 3 and standard deviation 6, putting the results in columns c1-c15. Note that we may consider the rows to constitute 100 samples of size 15 each. Determine the .9 level confidence intervals for the mean (which we know is 3) for each row, and determine if your results look reasonable — knowing you expect about 90 of these intervals to include the value 3.
   Hints: Use the *rmean*, *rstdev* and *let ci = ...* Minitab commands to help create adjacent columns with the lower and upper confidence limits.

2. Using the $1 - \alpha$ confidence interval for the variance, σ^2 of a sample of n independent identically distributed normal random variables,

$$\left(\frac{(n-1)s_n^2}{\chi_{n-1,\,1-\alpha/2}^2}, \frac{(n-1)s_n^2}{\chi_{n-1,\,\alpha/2}^2} \right)$$

   (from the answer to Problem 5.11.3. on page 157) where $\chi_{n-1,\,1-\alpha/2}^2$ is the upper $100(1 - \alpha/2)$ percentile of the *chi-square* distribution with n-1 degrees of freedom (see footnote1. on page 146) generate $n = 15$ normal random samples of size 100 each, with mean 3 and standard deviation 6, putting the results in columns c1-c15. Note that we may consider the rows to constitute 100 samples of size 15 each. Determine the .9 level confidence intervals for the standard deviation (which we know is 6) for each row, and determine if your results look reasonable — knowing you expect about 90 of these intervals to include the value 6.

3. Suppose that you obtained the following mileage values (in miles per gallon) from 10 full tanks of gasoline in your car
   15.0 19.1 16.7 17.8 17.5 17.9 16.2 19.6 15.1 16.9

   a. Making appropriate assumptions, obtain .99 level confidence intervals for your expected mileage and your standard deviation of mileage.
   Hints: The information in Exercises 1. and 2. above will be needed.

   b. Assuming your gas tank holds 20 gallons, how far can you safely go on one tank of gas.

---

1. Of course, there may be available information based on past experience (or a pilot project carried out to get an upper probability bound for the standard deviation [see Problem 5.11. above]) so that the situation may not be so uncontrollable.

> *Hint*: The Bonferroni inequalities (Theorem 4.8 on page 103) will be helpful.

Very useful confidence intervals for specified population percentiles can be gotten using various order statistics. We show how in the optional section to follow, and summarize immediately after.

⌐

We start by reproducing Equation 5.5 on page 147,

$$P\{X_{[j]} \le q\} = P\{S_n \ge j\} = 1 - P\{S_n \le j - 1\} \qquad 5.11$$

where $X_{[j]}$ is the j-th order statistics from the sequence $X_1, X_2, ..., X_n$ (see Definition 5.4 on page 147) and S_n has the binomial distribution with parameters n and $p = F(q) = P\{X_i \le q\}$.

To keep from complicating matters unduly, we will restrict consideration to the case where all population percentiles are unique, and population percentiles q_p and q_{p^*} are unequal whenever $p \ne p^*$. This will happen if there is a probability density which is never 0 over some interval which includes all possible values of the random variables described by this density.

Now, if we want to obtain a lower $1 - \alpha/2$ confidence limit for a specified $100p$ percentile, q_p, we try to choose the order statistic, $X_{[j]}$, with the largest index, j, such that

$$P\{X_{[j]} \le q_p\} \ge 1 - \alpha/2$$

From Equation 5.11 above, this is equivalent to choosing the largest integer, j, such that

$$1 - P\{S_n \le j - 1\} \ge 1 - \alpha/2 ,$$

i.e., the largest integer, j, such that

$$P\{S_n \le j - 1\} \le \alpha/2$$

where S_n has the binomial distribution with parameters n, p (so that j is a function of p). Restricting ourselves to the binomial distribution with parameters n and p we need only search for the largest integer value s such that the cumulative probability $P\{S_n \le s\}$
(which can be found using the Minitab menu bar
Calc -> Probability Distributions, -> Binomial -> Inverse Cumulative Probability)

satisfies $$P\{S_n \le s\} \le \alpha/2 .$$

Call this value $s_{\alpha/2}$ (note that $s_{\alpha/2}$ is a function of p). We now set $j - 1 = s_{\alpha/2}$, to obtain $j = s_{\alpha/2} + 1$.

The lower $1 - \alpha/2$ confidence limit for the $100p$ percentile, q_p is $X_{[j]}$ (recall Definition 5.3 on page 146).

The approach to finding an upper confidence limit for q_p is to note that because

$$X_{[j]} \leq q \text{ if and only if } S_n \geq j$$

(given at the start of this section) it follows that

$$X_{[j]} > q \text{ if and only if } S_n < j$$

where S_n is a binomial variable arising from trials whose probability of *success*,

$$P\{Y_i = 1\} = P\{X_i \leq q_p\} = F(q_p) = p.^1$$

Hence, to obtain an upper $1 - \alpha/2$ confidence limit for q_p, we look for the smallest integer s^* such that $P\{S_n \leq s^*\} \geq 1 - \alpha/2$. Call this $s^*_{1-\alpha/2}$. The upper confidence limit for q_p that we want is $X_{[j]}$ where $j = s^*_{1-\alpha/2} + 1$.

We summarize the results of the section above as follows.

Theorem 5.12: Confidence limits and intervals for percentiles

Suppose we have independent identically distributed random variables, $X_1,...,X_n$, with a density, completely specified by some (possibly vector) parameter, ϑ, which is either everywhere nonzero, or which is nonzero at each point in some interval (which may depend on ϑ) and 0 elsewhere. Let $X_{[1]}, X_{[2]},...,X_{[n]}$ denote the order statistics from $X_1,...,X_n$ (see Definition 5.3 on page 146). Let $s_{\alpha/2}$ denote the largest integer value s such that the cumulative probability $P\{S_n \leq s\}$ satisfies

$$P\{S_n \leq s\} \leq \alpha/2$$

where S_n has the binomial distribution with parameters n and p; and let $s^*_{1-\alpha/2}$ denote the smallest integer value, s^* such that

$$P\{S_n \leq s^*\} \geq 1 - \alpha/2$$

Then $\left[X_{[s_{\alpha/2} + 1]}; X_{[s^*_{1-\alpha/2} + 1]} \right]$ is a $1 - \alpha$ level confidence interval for the $100p$ percentile, q_p, of the distribution of the X_i

---

1. One has to be exceedingly careful here. The asserted result arises strictly from the fact that if $A = B$ then $A^c = B^c$. It might be tempting to conclude that because $X_{[j]} > q$ if and only if $S_n < j$ it might follow that $X_{[j]} \geq q$ if and only if $S_n \leq j$. This is false, as can be seen from the following diagram.

$$\underline{\hspace{2cm}|\underline{\hspace{0.3cm}!\hspace{0.3cm}}|\hspace{1cm}}\text{Here we see } X_{[j]} < q \text{ and } S_n = j. \text{ But}$$
$$\quad\quad X_{[j]} \quad q \quad X_{[j+1]}$$

if $X_{[j]} \geq q$ if and only if $S_n \leq j$, taking complements would yield

$X_{[j]} < q$ if and only if $S_n > j$ contradicting the above diagram's results.

$X_{[s_{\alpha/2}+1]}$ is a $1-\alpha/2$ level lower confidence limit for q_p

$X_{[s^*_{1-\alpha/2}+1]}$ is a $1-\alpha/2$ level upper confidence limit for q_p

❑

Practice Exercises 5.7

1. From a single random sample of 201 observations on the uniform distribution on the interval $[0;1]$ generate

   a. an approximate .95 level confidence interval for the mean based on the sample mean
      Hint: The standard deviation of such a uniform variable is $1/\sqrt{12}$.

   b. a .95 level confidence interval for the median (which is the same as the mean) based on the order statistics
      Hint: You can determine all you need to know about the distribution of the order statistics by getting a complete printout of the cumulative binomial distribution for $n = 201$ and $p = .5$, by first putting the integers from 0 to 200 in some column (use the *set* command).

   Which of the two estimates seems better? Explain.
   Hint: There is a good chance that the mean of the poorer estimate is closer to the true mean than that of the better one. So it wouldn't be convincing to choose as the better estimator, the one closer to the mean. Nonetheless, there is a choice which is well justified.

2. Generate 201 columns of 100 observations each from the uniform distribution. Using the *rmean* and *rmedian* commands put the row means in column 203 and the row medians in column 204. Use the describe command on column 203 and 204. Now decide which is the better estimator of the mean, the sample mean or the sample median.
   Hint: Both of these quantities have an approximately normal distribution, the sample mean directly from the Central Limit Theorem, and the sample median because the order statistics have a distribution that is essentially binomial (see Equation 5.12 on page 162) and hence which is approximately normal. (Save this dataset for future use.)

3. Same as previous question for a t distribution with 3 degrees of freedom. If you don't know the mean of this distribution, guess it from a histogram based on enough data.

Note that the methods of these last two questions (which involve very little theory) can be used to help decide which is the better of two location parameter estimators, both of which have at least approximately normal distributions.

5.5 Confidence Limits and Interval for Binomial p

A very important topic which is needed to yield precise results about the sample size needed in many Monte Carlo simulations, as well as for comparing the quality of different estimators using Monte Carlo methods, is that of confidence intervals for the parameter p in the binomial distribution; that is, confidence intervals for the probability, p, of success of an experiment which can either be a success or a failure.

To introduce our approach, we consider a statistical situation in which the parameter or interest completely describes the distribution of the random variables to be observed. This approach is due to Neyman, and the assumptions are

- independent identically distributed random variables $X_1, X_2,...,X_n$

- each distribution in the family being considered is completely determined by a scalar parameter, q, for which upper and lower confidence limits are desired

- there is an estimator $S = g(X_1, X_2,...,X_n)$ of q

- for each value of q, values $L_{1-\alpha/2}(q), U_{1-\alpha/2}(q)$ can be found, both of which are strictly continuous[1] and increasing as q increases,[2] satisfying

$$P\{S \geq L_{1-\alpha/2}(q)\} = 1 - \alpha/2 \quad \text{and} \quad P\{S \leq U_{1-\alpha/2}(q)\} = 1 - \alpha/2 \qquad 5.12$$

Because of the continuity and strict increase of the functions $L_{1-\alpha/2}$ and $U_{1-\alpha/2}$, solving for q, these probabilities can be written in the form

$$P\{q \leq L_{1-\alpha/2}^{-1}(S)\} = 1 - \alpha/2 \quad \text{and} \quad P\{U_{1-\alpha/2}^{-1}(S) \leq q\} = 1 - \alpha/2 \quad 5.13$$

So, using Definition 5.7 on page 152, we find that

$$U_{1-\alpha/2}^{-1}(S) \text{ is a lower exact } 1 - \alpha/2 \text{ confidence limit for } q$$

and

$$L_{1-\alpha/2}^{-1}(S) \text{ is an upper exact } 1 - \alpha/2 \text{ confidence limit for } q.$$

We show that for $0 < \alpha < 1$ the interval

$$[U_{1-\alpha/2}^{-1}(S); L_{1-\alpha/2}^{-1}(S)] \text{ is an exact } 1 - \alpha \text{ confidence interval for } q \text{ in}$$
this optional section that follows.

Because of the way the functions $L_{1-\alpha/2}$ and $U_{1-\alpha/2}$ were constructed, we see that $L_{1-\alpha/2} < U_{1-\alpha/2}$.

1. Meaning that as q increases by a small amount $L_{1-\alpha/2}(q)$ and $U_{1-\alpha/2}(q)$ vary by a small amount.
2. The case where the functions L and U are decreasing is handled similarly.

Examining Figure 5-2 below

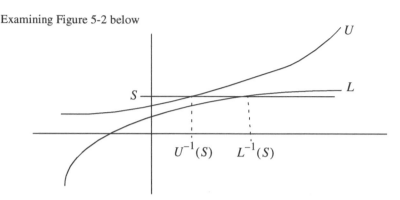

Figure 5-2 Illustrating Why $U_{1-\alpha/2}^{-1} < L_{1-\alpha/2}^{-1}$

it follows that $U_{1-\alpha/2}^{-1} < L_{1-\alpha/2}^{-1}$, so that $[U_{1-\alpha/2}^{-1}(S);L_{1-\alpha/2}^{-1}(S)]$ always evaluates to a nonempty interval.

In the following computation, we omit the subscripts $1-\alpha/2$ to keep the expressions manageable. We find, using DeMorgan's Law, Problem 4.5.1.d. on page 98, that

$$
\begin{aligned}
P\{U^{-1}(S) \le q \le L^{-1}(S)\} &= P(\{U^{-1}(S) \le q\} \cap \{q \le L^{-1}(S)\}) \\
&= 1 - P([\{U^{-1}(S) \le q\} \cap \{q \le L^{-1}(S)\}]^c) \\
&= 1 - P(\{U^{-1}(S) \le q\}^c \cup \{q \le L^{-1}(S)\}^c) \\
&= 1 - P(\{U^{-1}(S) > q\} \cup \{q > L^{-1}(S)\}) \\
&= 1 - P(\{S > U(q)\} \cup \{L(q) > S\}) \\
&= 1 - P(\{S > U(q)\} \cup \{S < L(q)\}) \\
&= 1 - P\{S > U(q)\} + P\{S < L(q)\} \\
&= 1 - (\alpha/2 + \alpha/2) \\
&= 1 - \alpha
\end{aligned}
$$

which shows this to be an exact $1-\alpha$ level confidence interval.

You might wonder why we put in so much detail here. The answer is that in very similar situation, the result does not hold. Namely, we might have a lower exact $1-\alpha/2$ level confidence limit for a normal population mean based on the sample mean, and an upper exact $1-\alpha/2$ level confidence limit for this same population mean based on the sample median, both based on the same sample of size $2n+1$. If we denote the lower confidence limit by $U^{-1}(X_{2n+1})$ and the upper confidence limit by $L^{-1}(X_{[n+1]})$, the events $\{U^{-1}(X_{2n+1}) \le \mu\}$ and $\{L^{-1}(X_{[n+1]}) \ge \mu\}^c$ need not be disjoint. So we must use the Bonferroni inequality (Theorem 4.8 on page 103), which does not yield the exact confidence.

If Equalities 5.12 on page 162 are modified to

$$P\{S \ge L_{1-\alpha/2}(q)\} \ge 1-\alpha/2 \quad \text{and} \quad P\{S \le U_{1-\alpha/2}(q)\} \ge 1-\alpha/2 \qquad 5.14$$

with both $L_{1-\alpha/2}(p)$ and $U_{1-\alpha/2}(p)$ strictly increasing with p and continuous (small changes in p yield small changes in these values) the conclusions (which yield slightly longer confidence intervals or poorer confidence limits) are:

for $0 < \alpha < 1$

$$U_{1-\alpha/2}^{-1}(S) \text{ is a lower } 1-\alpha/2 \text{ level confidence limit for } q \qquad 5.15$$

$$L_{1-\alpha/2}^{-1}(S) \text{ is an upper } 1-\alpha/2 \text{ level confidence limit for } q. \qquad 5.16$$

The interval

$$[U_{1-\alpha/2}^{-1}(S);L_{1-\alpha/2}^{-1}(S)] \text{ is a } 1-\alpha \text{ level confidence interval for } q. \quad 5.17$$

Neyman's approach to confidence limit construction can be extended to certain discrete distributions, such as the binomial and Poisson. The extension is extremely useful.   A minor difficulty arises from the fact that with discrete distributions it is not possible to find functions $L_{1-\alpha/2}(\ )$ and $U_{1-\alpha/2}(\ )$ satisfying Conditions 5.12 on page 162; for the binomial distribution we will be able to provide a complete solution by first finding the function $U_{1-\alpha/2}$ such that for each p with $0 < p < 1$, the value $U_{1-\alpha/2}(p)$ is the smallest integer such that

$$P_p\{S \le U_{1-\alpha/2}(p)\} \ge 1-\alpha/2 \qquad\qquad 5.18$$

From this it will be quite easy to find a lower $1-\alpha/2$ level confidence limit for p. Actually we will find two lower confidence limits — a lower $1-\alpha/2$ level confidence limit for p.   We will then find an upper $1-\alpha/2$ level confidence limit for p by finding a lower $1-\alpha/2$ level confidence limit for $1-p$ (using the same method).

First we establish that $U_{1-\alpha/2}(p)$ is nondecreasing as p increases; in fact, the function $U_{1-\alpha/2}$ is constant on intervals which include their right end points, (but not their left ones) as illustrated in Figure 5-3. Here is the reasoning.

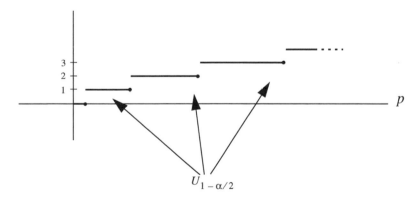

Figure 5-3 Illustrating Behavior of the Function $U_{1-\alpha/2}$

For $0 < p < 1$ and for each fixed integer s, when S_n is the number of successes in n independent trials, the quantity $P_p\{S_n \le s\}$ increases [decreases] by a small amount when p increases [decreases] a small amount.[1] So if

$$P_p\{S_n \le U_{1-\alpha/2}(p)\} > 1-\alpha/2 \qquad\qquad 5.19$$

for any p, for all nonzero h of sufficiently small magnitude (dependent on p), we have

---

1. This can be established from the formula for the binomial cumulative distribution function, derived in the answer to Problem 4.39.3.b on page 141.

$$P_{p+h}\{S_n \leq U_{1-\alpha/2}(p)\} > 1 - \alpha/2 \qquad\qquad 5.20$$

(because keeping nonzero h sufficiently near 0, means that the left side of 5.20 can be kept close enough to the left side of 5.19 to preserve the right hand inequality).

By the definition of $U_{1-\alpha/2}(p)$, we know that

$$P_p\{S_n \leq U_{1-\alpha/2}(p) - 1\} < 1 - \alpha/2$$

and from the previous arguments about small magnitude nonzero h, for sufficiently small h we also have

$$P_{p+h}\{S_n \leq U_{1-\alpha/2}(p) - 1\} < 1 - \alpha/2 .$$

This shows that $U_{1-\alpha/2}(p+h)$ cannot differ from $U_{1-\alpha/2}(p)$, proving that $U_{1-\alpha/2}(p)$ remains constant near any value of p for which

$$P_p\{S_n \leq U_{1-\alpha/2}(p)\} > 1 - \alpha/2 .$$

The only points of increase are those at which

$$P_p\{S_n \leq U_{1-\alpha/2}(p)\} = 1 - \alpha/2$$

at which $U_{1-\alpha/2}(p)$ jumps up one unit with any sufficiently small increase in p. This establishes the validity of Figure 5-3 on page 164.

To convert to the situation described by Neyman for constructing confidence limits and intervals for the binomial distribution, we construct a strictly increasing function, $\tilde{U}_{1-\alpha/2}$ which tries to *hug* $U_{1-\alpha/2}$ and does not lie below $U_{1-\alpha/2}$ as shown in Figure 5-4.

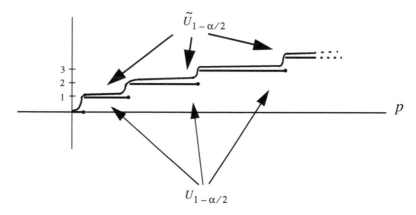

Figure 5-4 Illustrating behavior of the functions $U_{1-\alpha/2}$ and $\tilde{U}_{1-\alpha/2}$

The modified Neyman construction (5.14, 5.15, 5.16, and 5.17 on page 164) can be applied to the function $\tilde{U}_{1-\alpha/2}$. What it shows is that a lower $1 - \alpha/2$ level confidence limit for the binomial parameter p is

$$\tilde{U}_{1-\alpha/2}^{-1}(S_n) \quad \text{whose value is} \quad \tilde{U}_{1-\alpha/2}^{-1}(S_{obs}).$$

As can be seen from Figure 5-4, **this lower** $1 - \alpha/2$ **confidence limit for the binomial** p **is the value of** p **for which**

$$P_p\{S_n \leq S_{obs} - 1\} = 1 - \alpha/2 \qquad\qquad 5.21$$

where S_{obs} is the actual observed number of successes, and S_n has the binomial distribution with parameters n and p.

As mentioned earlier, we will obtain an upper confidence limit for p from a lower $1 - \alpha/2$ level confidence limit for $1 - p$, as follows. Let N_F be a random variable representing the number of failures in n independent trials, each of which has probability $1 - p$ of failure, and let N_{obs} be the actual observed number of failures in these trials. In the following expression the subscript of P indicates the unknown parameter which determines the binomial distribution of the random variable in the argument to the probability distribution. From the result leading to Equation 5.21 we see that a lower $1 - \alpha/2$ level confidence limit for $1 - p$ is the value of $1 - p$ for which

$$P_{1-p}\{N_F \leq N_{obs} - 1\} = 1 - \alpha/2$$

Using the fact that $N_F = n - S_n$ and $N_{obs} = n - S_{obs}$, substituting these into the equation above, we find that the upper $1 - \alpha/2$ level confidence interval for p is the value of p for which

$$P_p\{n - S_n \leq n - S_{obs} - 1\} = 1 - \alpha/2 \text{ or } P_p\{S_n \geq S_{obs} + 1\} = 1 - \alpha/2$$

or, finally, the desired upper confidence limit for p is the value of p for which

$$P_p\{S_n \leq S_{obs}\} = \alpha/2. \qquad\qquad 5.22$$

We summarize the main result of the optional material just concluded in the following theorem.

Theorem 5.13: Binomial confidence limits and intervals

Suppose the random variable S_n represents the number of successes in n independent trials, each of which has (unknown) probability of success p, and S_{obs} denotes the actual observed number of successes in these trials.

A lower $1 - \alpha/2$ level confidence limit, $p_{L, 1-\alpha/2}$, for p, is the value of p for which

$$P_p\{S_n \leq S_{obs} - 1\} = 1 - \alpha/2$$

An upper $1 - \alpha/2$ level confidence limit, $p_{U, 1-\alpha/2}$, for p is the value of p for which

$$P_p\{S_n \leq S_{obs}\} = \alpha/2$$

A $1 - \alpha$ confidence interval for p is the interval

$$[p_{L, 1-\alpha/2}; p_{U, 1-\alpha/2}].$$

❑

Example 5.8: Confidence interval for p when $n = 100$

Suppose 100 independent success-failure trials, each with unknown probability p of success are run, with 44 successes and 56 failures resulting. Find a .95 level confidence interval for p.

To aid in this search, the following macro, *climsrch.mac*, whose use is self-explanatory, is presented. Here is the code.

```
GMACRO

climsrch

note enter n Sobs and p0 (first guess for binomial confidence limit)

 # If note is the first word on some line, material to its right is printed on the screen.

note If p0 > S/n we search for upper confidence limit


read c1 c2 c3;

file "TERMINAL"; # to read from a terminal

NOBS 1.

# the data is read in


copy c1 k1

copy c2 k2

copy c3 k3

# the data is copied to constant values for comparisons and arithmetic operations


let k4 = k2/k1

# computation of Sobs to compare with trial p for deciding

# whether upper or lower confidence limit is to be sought


if k3 < k4

 let k7 = 0

 let k5 = k2 -1 # lower confidence limit   S -> S-1

else

 let k7 = 1

 let k5 = k2 # upper confidence limit

endif


cdf k5;

binomial k1 k3.

# compute binomial cdf value for given Sobs and trial p


do k6 = 1:50 # repeat process for different trial values of p
```

if k7 = 0

note enter next trial p for lower conf lim or terminate with END

note for lower conf lim we're looking for high probability 1 - alpha

note the lower the chosen p the higher the resulting 1 - alpha

else

note enter next trial p for upper conf lim or terminate with End

note for upper conf lim we're looking for low probability alpha

note the higher the chosen p the lower the resulting alpha

endif

read c3;

file "TERMINAL";

NOBS 1.

copy c3 k3

cdf k5;

binomial k1 k3.

enddo

ENDMACRO

The macro *clmsrch.mac* lets you interactively search for binomial confidence limits. A program, *binclim.exe*, which runs on a PC, whose inputs are the sample size, observed proportion of 1s (successes), and desired confidence, and whose output is the desired confidence interval, will be available on the CRC website http://www.crcpress.com.

■

Practice Exercises 5.9

1. My brother-in-law invented a device to age wine in glass bottles, using a wooden dowel with carefully engineered oxygen-transfer properties. When he visited wine shops, out of 20 people that he met, 18 of them wanted to buy his device on the spot. He figured that he should be able to garner about 10% of the wine-store-patron market for his invention. Was he being realistic?

    Hint: Find a relevant confidence limit for the probability that a randomly selected person from this market would buy this device.

2. Compare the approximate .95 level confidence interval for binomial p, $\hat{p} \pm 1.96\sqrt{\hat{p}(1-\hat{p})/150}$ (where $\hat{p} = S_{150}/150$) with the .95 level confidence interval for p using Theorem 5.13 on page 166 when the sample size $n = 150$, using the *climsrch* macro on page 167, in the following cases.

    a. $S_{obs} = 148$

b. $S_{obs} = 145$

c. $S_{obs} = 125$

3. Find an upper .95 level confidence interval for the probability of success of an experiment which had 25 successes in 25 trials. What faith would this have given you, in advance, of the chances of success of the 26-th Spaceship Challenger, assuming that there were no basic changes made in the Challenger's structure?

4. A pair of dice is tossed 200 times, and comes up even in 130 of these tosses. Do you have any reason to be suspicious?

Problem 5.14

Referring back to the data generated in Exercise 5.7.2. on page 161, generate column c205 to contain the row maximum of columns c1-c201, c206 to contain the row minimum, and c207 to contain the average of columns c205 and c206. Determine what proportion of intervals centered at the values in column c207, and whose lengths are 2 x 1.96 x sqrt(1/[12 x 201]) include the true mean (of .5). Note, these are the lengths of the .95 level confidence intervals for this mean, based on the sample mean. It will be significant if this proportion exceeds .95 substantially.

Hint: To compute what is needed, the techniques introduced in Example 1.3 starting on page 10 and concluding on page 14 might be used. Alternatively, histograms of columns c203 and c208 give a rough idea of the answer being sought.

From this proportion, what can you say about the confidence of coverage that this interval has? (You should use the result on binomial confidence limits, Theorem 5.13 on page 166; the macro, *climsrch.mac* on page 167 might also be helpful.)

5.6 Comparing Estimators

Current scientific research can be very expensive, including the gathering of data. Just the cost of getting 5 more observations (one per mouse) might run as high as $400. So there is great interest in getting the best estimates you can from your data.

There are situations in which you may have a reasonable choice between two estimators, and want to know which one is better. You should realize that there is a good chance that neither is better — i.e., that the choice is not one that can be reasonably made. For instance, if you want to estimate some parameter ϑ, and have two estimators, $g(X_1,...,X_n)$ and $h(X_1,...,X_n)$, it can well be that $g(X_1,...,X_n)$ tends to be closer to ϑ than $h(X_1,...,X_n)$ for some values of ϑ while $h(X_1,...,X_n)$ tends to be closer to ϑ than $h(X_1,...,X_n)$ for other values of ϑ.

For a meaningful example, maybe g is a better estimate of the cost of living, ϑ, during a recession (recession being associated with a low living cost) while h is a better estimate of ϑ during an inflation (inflation indicating a higher living cost).

It may be the case that for some particular value, ϑ_0 of ϑ we have

$$P_{\vartheta_0}\{|g(X_1,...,X_n)| < c\} > P_{\vartheta_0}\{|h(X_1,...,X_n)| < c\}$$

for some values of c, while for other values of c we have

$$P_{\vartheta_0}\{|g(X_1,...,X_n)| < c\} > P_{\vartheta_0}\{|h(X_1,...,X_n)| < c\}.$$

So it seems that generally, even the problem of finding the better of two estimators has no solution. Nonetheless, there are sufficiently many important cases in which there is a solution, to make it worthwhile to examine some of them. Here are a few illustrations.

Example 5.10: Mean versus median — the standard deviation approach

Suppose we are using a measuring instrument (such as a device for measuring blood pressure) whose readings consist of independent normal random variables with unknown mean μ, and standard deviation σ (possibly unknown). The mean and median of a normal distribution are the same. Two contending estimates for μ are therefore the sample mean and the sample median. The sample mean has a normal distribution (see Theorem 4.40 on page 141) and from Theorem 5.5 on page 147 we know that the distribution function of the sample median can be well approximated by a normal distribution function for large sample size. So you would think that in the problem being considered, at least for sufficiently large sample size, the estimator with the smaller standard deviation is more likely to be within any specific distance c of μ than the one with the larger standard deviation. That is to say, the estimator with the smaller standard deviation would seem better.[1] Even knowing that there may be subtle problems with this approach, we will examine the two estimators in question from this point of view via Monte Carlo simulation.

To carry this out, we first generate 51 columns each with 1000 rows of independent standard normal data, via the Minitab line command

        random 1000 c1-c51<Enter>

standard normal data being adequate for our purposes. Then we generate the sample means and sample medians of the 1000 rows via the commands

                    rmean c1-c51 c53 <Enter>

                    rmedian c1-c51 c54 <Enter>

Next we type   describe c53-c54 <Enter>

---

1. Actually, it is not quite that simple, because, for any given n, no matter how large, it is conceivable that the Central Limit Theorem has not yet taken hold well enough to conclude that the estimator with the smaller standard deviation is more likely to be within a specific distance, c, of μ than the one with the larger standard deviation.

from which we obtain the following output:

Descriptive Statistics: C53, C54

Variable	N	Mean	Median	TrMean	StDev	SE Mean
C53	1000	-0.00026	-0.00459	-0.00102	0.13779	0.00436
C54	1000	0.00341	0.00480	0.00351	0.17310	0.00547

Variable	Minimum	Maximum	Q1	Q3
C53	-0.44482	0.45150	-0.09330	0.08959
C54	-0.64119	0.55235	-0.12125	0.12270

So it seems that the standard deviation of the sample mean (column c53) is smaller than that of the sample median, which would indicate that in the case of normal data, the sample mean is a better estimator than the sample median. We could nail this down even more firmly by obtaining confidence intervals for the given standard deviations. Due to the space it would take, we will not do this, but rather, take a more direct path in the next example.

∎

The approach to be taken in the next example is suggested by the two histograms of the data of columns c53 and c54 in the example just presented.

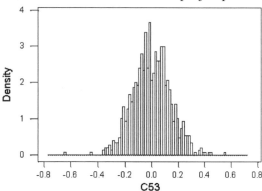

Figure 5-5 Histogram of Sample Means

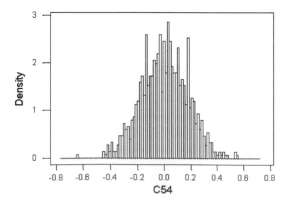

Figure 5-6 Histogram of Sample Medians

Example 5.11: Normal sample mean versus sample median — direct approach

Looking at these histograms, we expect that the sample means will tend to be closer to the desired mean/median than the sample medians. To check this out we sort that sample mean and sample median columns, using the Minitab commands

sort c53 c56 <Enter>

sort c54 c57 <Enter>

For any constant $c > 0$, we can obtain estimates of

$$P\{|\bar{X} - 0| \le c\} = P\{|\bar{X} - E(X)| \le c\}$$

and

$$P\{|X_{[51]} - 0| < c\} = P\{|X_{[51]} - Med(X)| \le c\},$$

just by counting the number of sample means and sample medians, respectively, within a distance of c units from 0. These two random variables have the binomial distribution with parameters $n = 101$ and

$$p_{mean,c} = P\{|\bar{X} - E(X)| \le c\}, \quad p_{med,c} = P\{|X_{[51]} - Med(X)| \le c\}.$$

In our simulation, we found that for $c = .23$ the .95 level confidence intervals were

for $p_{mean,c}$ (.882; .920)   for $p_{med,c}$ (.795; 843)

leading us to conclude (as suggested by the histograms above) that the sample mean is more likely to be closer than .23 to the mean/median than is the sample median.

∎

Practice Exercises 5.12

1. Using the methods developed here, compare the sample mean with the sample median as estimates of the mean/median for the following distributions:

    a. Any uniform distribution

    b. The t distribution with n degrees of freedom, $n = 2,7,12$

2. For any uniform distribution, as estimates of the mean, compare the average of the largest and smallest values with the sample mean.

Problems 5.15

1. For any normal distribution, as estimates of the standard deviation, compare the sample standard deviation with a suitable multiple of the interquartile range. (To get this multiple, you might use a Monte Carlo approach. That is, estimate the interquartile range, and then find out what you must multiply it by to get the value σ, of the normal distribution you sampled from. Check that this value does not depend on σ.)

2. Try to determine for the normal distribution which of the interpercentile ranges (suitably normalized) gives the best estimate of the standard deviation. (I haven't worked out the answer as of the time this was written, but the approach should be fairly clear.)

3. For the Cauchy distribution, see if you can determine which of the interpercentile ranges (suitably normalized) seems to give the best estimate of the interquartile range.

5.7 The Bootstrap

Suppose you did not know the expression $\sqrt{np(1-p)}$ for the standard deviation of a binomial random variable with parameters n and p, but wanted somehow to determine it. You might try generating many binomial variables, $S^{(1)}, S^{(2)},...,S^{(m)}$, with these parameters, and use the sample standard deviation

$$\sqrt{\frac{1}{m-1} \sum_{i=1}^{m} (S^{(i)} - \bar{S})^2}$$

(where $\bar{S} = \frac{1}{m} \sum_{i=1}^{m} S^{(i)}$), as an estimate of the standard deviation you want.

But what if all you had was the sequence of values of $X_1, X_2,...,X_n$, where X_i is a random variable giving the number of successes on trial i. (Remember that X_i is either 1 (with probability p) or 0 (with probability $1 - p$) where p is unknown to you. Is there any reasonable way **to determine the standard deviation of the sum**

$$S = \sum_{i=1}^{n} X_i$$

of your observed data with just the given information?

The answer to this question is *yes*, provided that n is reasonably large. Here is why. If the sample is reasonably large, the value of S/n is likely to be close to p. But p completely determines the joint distribution of the statistically independent $X_1, X_2,...,X_n$. If we repeatedly independently sample from the sequence of observed values, $x_1, x_2,...,x_n$, of $X_1, X_2,...,X_n$, we will be sampling from a population in which the probability of a 1 is a value, $\tilde{p}$ which in most cases will be close to p. If the standard deviation of the sum,

$$S^* = \sum_{i=1}^{n} X_i^*$$

of independent random variables, X_i^*, from this population, was not likely to be close to the standard deviation of S, this would mean that random variables whose descriptions are very close could have very different standard deviations. Since we cannot expect a model to be a perfect description of a system it is used to describe, most models we derive would be useless for prediction. But we know from practical experience that many imperfect models have proved quite useful for this purpose. So, we would expect the standard deviation of S^* to be close to that of S. Now we would expect the sample standard deviation of sums, S_j, of sequences of observed values, $x_{1j}, x_{2j},...,x_{nj}$ drawn independently and with replacement from the sequence $x_1, x_2,...,x_n$ of observed values of $X_1, X_2,...,X_n$ to be close to the standard deviation of S^*, and hence, close to the standard deviation of S that we are seeking.

This argument does not prove our claim, but I hope it makes it seem reasonable. What is necessary if we want to use this approach, is to determine when it can be expected to give us adequate results, and when it is not trustworthy. We will return to this issue shortly.

Topic 5.16: Summary of bootstrap for binomial standard deviation

Suppose we have observed the values, $x_1, x_2,...,x_n$, of independent Bernoulli random variables $X_1, X_2,...,X_n$, variables with $P\{X_i = 1\} = p = 1 - P\{X_i = 0\}$, and want to estimate the standard deviation of the binomial sum

$$S = \sum_{i=1}^{n} X_i$$

solely from the observed values $x_1, x_2,...,x_n$, without any use of the formula relating this standard deviation to the value of p. The following is suggested.

Repeatedly and independently sample n observations **with replacement** from the sequence of observed values, $x_1, x_2,...,x_n$ of $X_1, X_2,...,X_n$. Call these samples $x_{1j}, x_{2j},...,x_{nj}$, $j = 1, 2,...,K$. Let S_j be the sum of this sequence, i.e.,

$$S_j = \sum_{i=1}^{n} x_{ij}$$

and let

$$\bar{S} = \frac{1}{n} \sum_{j=1}^{K} S_j.$$

For suitably large n and K, we would expect the sample standard deviation

$$\sqrt{\frac{1}{K-1} \sum_{j=1}^{K} (S_j - \bar{S})^2}$$

to be close to the standard deviation of S.

The generation of the sequences $x_{1j}, x_{2j},...,x_{nj}$ by sampling with replacement from the values $x_1, x_2,...,x_n$ of $X_1, X_2,...,X_n$ is called **resampling of** $X_1, X_2,...,X_n$.

This is the idea behind the *bootstrap* method. In this illustration the method is not needed, because we know a formula connecting n and p with the standard deviation of S. However, in many situations we estimate some quantity from a single data set, and would like to estimate the standard deviation of our quantity's estimate, and have no formula relating the value of the quantity to its standard deviation. The method just presented can be extended to handle this task, so long as the raw data from which the estimate of the quantity was formed is still available.

Let's see how well the suggested procedure works on our introductory example. In order to carry out the resampling computation for the case being considered, a macro, *samprep.mac* was written. It is reproduced below, and it computes the sample standard deviation that is described above.

```
GMACRO

samprep

erase c1-c1000  # to get a clean-looking spreadsheet

note input p, number rows and number of columns to be generated

      # what follows the word note is printed to the terminal screen

read c4 c5 c6; # This and the next command causes a pause for

              # terminal input of 3 values

file "TERMINAL";

NOBS 1.  # These last 3 lines read in 3 numbers typed in without an <Enter> and

              # continue on automatically

copy c4 k2  # This and the next two commands take the inputs and stores them

              # in 3 constants

copy c5 k3

copy c6 k4

random k3 c1; # generates original column from which to sample

bernoulli k2.

let k4 = k4 + 7  # to keep columns arising from resampling separated from input data

do k1=8:k4    # This loop repeatedly samples with replacement from column c1

              # produces columns c8 to ck4 of same length as column c1

sample k3 c1 ck1;

replace.

enddo

let k5 = k4 + 2 #create column number to store column sums of cols c8 to ck4
```

do k1 = 8:k4 # Loop to do actual summing of columns and storage of sums in col ck5

let k6 = k1 - 7

let ck5(k6) = sum(ck1) # puts binomial variables in column ck5

enddo

let c2 = ck5 # Now c2 has the binomial variables that are generated

stdev c2 # to calculate the sample sd of this sequence of binom variables

ENDMACRO

To run it, if. e.g., you copy it into the root directory, C:\, then you would type

%C:\samprep

I had the macro file in the directory C:\stbk\macs, so this was what was printed
after I typed %C:\stbk\macs\samprep

```
Executing from file: c:\stbk\macs\samprep.MAC
input p, number rows and number of columns to be generated
DATA> |
```

The following data was entered, with the indicated results

```
Executing from file: c:\stbk\macs\samprep.MAC
input p, number rows and number of columns to be generated
DATA> .3 200 600
          1 rows read.
```

Standard Deviation of C2

```
Standard deviation of C2 = 6.4581
```

The theoretical standard deviation of a **sum** of 200 Bernoulli variables, X_i, each of
whose probability of equalling 1 is $p = .3$, is $\sqrt{200 \times 0.3 \times 0.7} = 6.4807$. This is the value
we would like to estimate closely using the bootstrap.

The sample standard deviation of the values in column c1 (the original one
generated from 200 success-failure trials, with probability $p = .3$ of success) was
.4604, which is a pretty good estimate of the true standard deviation
$\sqrt{0.3 \times 0.7} = \sqrt{0.21} = 0.4583$. From the theory relating standard deviation of a
sum of variables with a given standard deviation, we know that the theoretically
based estimate of the standard deviation of the sum of the random variables
$X_1, X_2, ..., X_n$ is

$$\sqrt{200} s_{200} = \sqrt{200} \times 0.4604 = 6.511$$

which is the value the bootstrap is really estimating directly (where s_{200} is the
sample standard deviation of the observed X_i).

The bootstrap estimate gotten by taking 600 samples, each of size 200 with
replacement from the original column, is 6.4581.

So what we have is

true standard deviation of $\sum\limits_{i=1}^{200} X_i$ is 6.4807

estimate of standard deviation of each of these X_i via the sample standard
deviation is .4604

which leads theoretically to

an estimate of the standard deviation of $\sum\limits_{i=1}^{200} X_i$ with value 6.511

(which is close to the desired 6.4807)

The bootstrap estimate obtained from resampling from $x_1, x_2,...,x_n$ (and which
therefore should be close to 6.411) is 6.4581.

So the bootstrap estimate of the binomial standard deviation, based solely on the
data which provided the estimate of p, is likely to be close to something which is
likely to be close (when the sample size leading the estimate of p is large) to the
standard deviation of

$$\sum_{i=1}^{n} X_i$$

True value 6.4807, bootstrap estimate 6.4581.

Here is a case in which the bootstrap technique is really needed. Only a single
(possibly large) sample of ordered pairs, (x_i, y_i) is available, and from it the sample
correlation coefficient based on these pairs is formed (see Definition 3.19 on page
76). We want to estimate the standard deviation of this sample correlation coefficient,
and have no simple formula relating the standard deviation of the correlation
coefficient to its value.

Nonetheless, sampling from the data on which this estimate is based (assuming it
is a large enough sequence) should yield *resampled data* whose distribution is close
to that of the distribution from which the original data was drawn — because from the
frequency interpretation of probability, a small base interval (two-dimensional)
histogram of the original data set should be close to the (two-dimensional) probability
density function from which the original data was drawn. We would expect that the
correlation coefficient would vary continuously with the probability density and
probability functions (i.e., closeness of two such functions would yield close
correlation coefficients). Hence we would expect the sample correlation coefficients
generated by resampling from the data which generated the original correlation
coefficient would resemble a sample of correlation coefficients from the original
distribution. We could then use the sample standard deviation of these resampled
correlation coefficients to be close to the theoretical standard deviation of the original
sample correlation coefficient.

The only extra difficulty in writing a bootstrapping routine to estimate the standard deviation of the sample correlation coefficient is the bookkeeping, since resampling from a pair of columns can be done by repeated use of the Minitab command

sample k3 c1-c2 ck1-ck2;

replace.

embedded in a do-loop where k1 increases by 2 with each iteration, and k2 is always one unit larger than k1.

Practice Exercises 5.13

1. Try repeating the bootstrap estimate of the standard deviation of $\sum_{i=1}^{n} X_i$ for other values of n and p using the macro *samprep.mac*.

2. Repeatedly generate a large number, n, of pairs of variables, (X_i, Y_i) where the X_i are independent standard normal and $Y_i = X_i + Z_i$ where the Z_i are standard normal and independent of the X_i. Compute the many sample correlation coefficients between the X_i and Y_i, and use these sample correlation coefficients to estimate the standard deviation of these sample correlation coefficients (the complete Monte Carlo approach). If you plan to do Problem 5.17 use the modified bootstrap macro you will write there to estimate this sample correlation coefficient standard deviation, and compare results.

Problem 5.17

Modify the macro *samprep.mac*, so it can handle estimating the standard deviation of the sample correlation coefficient.

While the bootstrap technique allows the construction of good estimates in cases where theoretical methods are lacking, caution has to be exercised, since many hazards are likely to befall those who use it mindlessly. To see one of them, try using it to estimate the nonexistent standard deviation of the mean of a Cauchy random variable.

To get an idea of the sample size needed for a successful bootstrap estimate in a real situation, it is wise to make comparison with known solutions for distributions close to the type from which the data is coming. Especially in the cases where the bootstrap method is needed, if possible, do a full Monte Carlo procedure of generating the statistic whose standard deviation is desired — i.e., resample from the *true distribution* and see how well resampling from a data set from this distribution compares.

For more information about bootstrap techniques the book by Efron (who developed the original idea of the bootstrap) and Tibshirani [3] is recommended. The book by Chernick, [2], has an extensive bibliography of both theoretical and practical works on this subject.

Problem 5.18

When does it appear that sample percentiles are favored for learning about a population over sample means and sample standard deviations?

Hints: Generate one or more data sets of many independent identically distributed random variables, such as from standard normal, uniform, exponential or Poisson distributions. To such data sets append one very very large positive outlier and its negative. Now examine these data sets.

Review Exercises 5.14

1. What is the basic device used to get a handle on the distribution of the order statistics and the sample percentiles?

2. Suppose that independent identically distributed random variables $X_1, X_2, ..., X_{200}$ are normal with mean 0 and standard deviation 1.

   a. Find $P\{X_{[50]} < -1\}$.

   b. Find $P\{-2 < X_{[50]} < -1\}$.

3. Same as previous question but for Cauchy random variables with median equal to 0.

4. Why is so much attention paid to confidence intervals for the binomial parameter p?

5. Present some examples of the practical use of confidence intervals.

6. What price is paid by not knowing the standard deviation when obtaining a confidence interval for the mean of normal random variables?

7. Where might it be very useful to have an upper confidence limit for the standard deviation of a normal random variable?

8. For what important application is a confidence interval for the ratio of two standard deviations needed?

9. What is the basic problem addressed by the *bootstrap* method?

Chapter 6

Testing Hypotheses

6.1 Introduction

Many scientific investigations begin by asking the question of whether or not one or more specified quantities are related to some other specified quantity. This was certainly so in the following cases.

- Does cigarette smoking tend to cause lung cancer?
- Does a steady regimen of exercise tend to lead to longer life?[1]
- Is lengthy exposure to the sun bad for you?
- Is the new manufacturing process better than the current one?

In each of these situations a *yes-no* question is asked. Some other yes-no questions which have come up are

- Classify each cell in some sample as either *cancerous* or *noncancerous*
- Determine whether or not Bush will win the next presidential election.
- Decide whether or not we are in for an economic recession.
- Does lowering the speed limit save lives or resources?
- Does lowering the speed limit have any effect at all in Texas?
- Does the transponder data from the other aircraft indicate that a course change is needed to avoid a collision?

A procedure to furnish an answer to a *yes-no* question based on statistical data is called a **test of hypotheses**. Recall that the one thread common to all **statistical problems** is the assumption that our knowledge of the distribution governing the observed data is incomplete. Phrased more exactly, we do not know which probability distribution(s), from a class containing at least two distinct probability distributions, adequately describes our data. In testing statistical hypotheses, we have the following structure.

- some data gathered to help find a convincing answer some *yes-no* question we have in mind

- a class, P, of probability distributions, at least one of which is assumed to contain an acceptable statistical description of our data

- two exclusive and exhaustive subsets H_0, H_1 of P, one of which, say H_0, consists of those distributions in the class P corresponding to an answer of *yes* to our question and the other subset, H_1, corresponds to those distributions in P which would lead to an answer of *no*.

---

1. Of course it does, if you can keep it up long enough.

For historical reasons, H_0 is called the *null hypothesis* and in the medical and agricultural area usually corresponds to a situation of *no change* in the effect of a new treatment compared to the old standard, or of no change in the characteristics of some new strain compared to the one in current use. The hypothesis H_1 is called the alternative hypothesis, This terminology (and the emphasis on choosing H_0 unless there is compelling evidence to the contrary) arises from a conservative approach designed to avoid costly and often unnecessary operational shifts.

Definition 6.1: Test of hypotheses

A **test of hypotheses consists of** a breakup of the class *P of possible probability distributions of the data* into

the two disjoint and exhaustive subsets H_0, H_1

and

a breakup of the sample space into two disjoint and exhaustive subsets, one corresponding to *nonrejection of the hypothesis* H_0 and the other corresponding to *rejection of the hypothesis* H_0 *in favor of the hypothesis*

H_1. (Rejection of H_0 when the true distribution is in H_0 is called a *type I error*. Rejecting H_1 when the true distribution is in H_0 is a *type II error.*)

All of this may seem a bit abstract, but with luck the fog will clear some time after we have gone through the examples to follow.

Example 6.1: Are the airplanes going to collide?

When the commercial jet age arrived around 1957, one of the big problems was how to avoid midair collisions. The reason for this worry was that planes could be routinely approaching each other at speeds near 1000 miles per hour (somewhat over 1/4 of a mile per second). By the time one of the pilots might catch sight of the other airplane, and have enough data to decide that a course change was needed, it would likely be too late to avoid catastrophe. To complicate matters, the measured data would almost certainly fall short of perfection, which made the problem a statistical one.

We will concentrate on the problem of whether the current course of the airplanes will lead to a collision. The choice of which measurements to use is a very critical one in most problems, one which is quite dependent on the assumptions that are made. In this problem it is assumed that both airplanes are flying with a constant speed and direction. Measurements can be made on the distance between the airplanes, and if there were no error, they could be used to devise a collision avoidance algorithm. However, it turns out that even small errors in distance measurement would lead to unacceptable results if no other data were available. So, in addition to the distances, the bearing angles, BA1, BA2,... indicated in Figure 6-1, would also be measured.

The procedure to devise a test using the measurements above is based on the idea that under the constant velocity assumption, a collision of the center points of the two aircraft will be avoided if and only if the successive bearing angle measurements are changing, as illustrated in Figure 6-1.

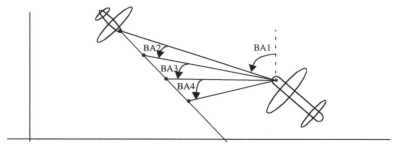

Figure 6-1 Illustrating Changing Bearing Angles on Non-collision Course

Keep in mind that the data to be gathered is the sequence of pairs (di, BAi) i = 1,2,... where di is the i-th distance measurement and BAi is the i-th bearing angle measurement. Assuming that the measurement error is due to small deviations from a constant velocity course and small deviations in distance and angle measurements, a reasonable first assumption is that the distance measurement errors are normal with mean 0 and standard deviation σd feet, the bearing angle measurements are normal, with mean 0 and standard deviation σa radians, and all errors are statistically independent of each other. The values of these standard deviations were supplied by engineers with a great deal of experience with the distance and angle measuring devices, and with the deviations from a constant velocity course that airplanes exhibit when under control of an autopilot. Having simplified the actual analysis somewhat, the probability distribution of the measured data can now be readily determined for any pair of constant velocity aircraft courses.

The procedure for collision warning is broken up into two parts. A smallest *miss distance* of *fm* feet is chosen, where the value of *fm* is chosen, so that a change in course of the type specified would actually result in a collision. The simplified question is *which of the two miss distances, fm or 0 is a better description of the data*? If 0 is a better description, then a collision avoidance maneuver is to be carried out. If *fm* is a better description, then no such maneuver is to be implemented. By *better description*, we will mean that the data is subjected to a sample space breakup called a *likelihood ratio test*[1] of the hypothesis H_0 that the data is governed by the distribution corresponding to a miss distance of 0, versus the alternative H_1 that the data is governed by the distribution corresponding to a miss distance of *fm* feet.

---

1. The *likelihood ratio (LR) test* is based on the value at the observed data of the ratio of the probability densities associated with the two chosen *miss distances, fm* and 0. It seems reasonable when $dat = (d1, BA1, d2, BA2,...,dn, BAn)$ denotes the observed data, to reject the collision hypothesis if $f_{dist = fm}(dat)$ exceeds some suitably chosen multiple of $f_{dist = 0}(dat)$, i.e., for large enough values of *LR*. The *LR* test is the breakup of the sample space referred to in Definition 6.1 .

Finally, the probability of a collision warning will be determined for all other possible minimum miss distances, in order to see whether the derived test procedure is acceptable.

The above discussion is a non-detailed outline of how a test of hypotheses was actually developed for aircraft collision warning.[1]

■

This first example illustrated the application of testing hypothesis methods to a complicated engineering problem, one which took a number of months to formulate and solve. More commonly, because of variability in the data produced in engineering and science, it is necessary to formulate many problems in the testing hypothesis framework. Here is a typical illustration.

Example 6.2: Do young mice react differently than aged ones?

At the University of Texas Medical Branch a large group of physiologists are studying the process of aging. One of the most common questions that comes up is whether some given treatment affects young mice differently from aged ones. If a random sample was chosen from the population of **all young mice**, and another random sample was picked from the population of **all old mice**, the variability among the mice would lead to the requirement of very large sample sizes in order to detect a real difference between the average reactions of young and old mice. For this reason experimentation is carried out on young and old mice from a strain of mice which was bred to be very uniform in their characteristics — this uniformity making it easier to detect differences in the *young versus old* reactions.[2] At this point the appropriate statistical procedure to use will depend on what assumptions the experimenter is prepared to make about the nature of the change to be expected, in the event that there is a change. This is where attention must be paid to the assumptions associated with the various commonly used test procedures.

For the purposes of this discussion we will assume that there are two groups of mice, one chosen at random from the young mice of the uniform population that has been chosen, and the other from the old mice of this population. The data has already been chosen and gathered by the experimenters. It is assumed that the distribution of the reaction to the treatment is the same for all the young mice, and the same for all of the aged mice, and that the observations made are statistically independent. The null hypothesis, H_0, corresponds to the distributions indicating no difference (in some sense) in the reactions of the young and aged mice. Four of the most common distinct assumptions that could then be made about any change are

1. The distribution of the characteristic being measured to detect a change in reaction to some treatment is normal, with mean value m_y for the young mice, m_a for the aged mice and unknown standard deviation σ. (The

---

1. For more details, see the paper by Rosenblatt [8].
2. Of course, this restriction to a single strain might make any generalization to human populations less reliable, so whether it is the right approach is questionable.

measurement might just be the difference within each mouse of the value of the characteristic before and after some treatment.)

Conventionally, the null hypothesis, H_0, corresponds to those pairs of normal distributions $(N(m_y, \sigma), N(m_a, \sigma))$ with $m_y = m_a$, and the alternative hypothesis, H_1, those with $m_y \neq m_a$. Actually, the experimenters really do not want to accept the alternative hypothesis, H_1 unless m_y is sufficiently different from m_a. We will worry later about how this difficulty is handled.

2. The distribution of the characteristic being measured to detect a change in reaction to some treatment is normal, with mean value m_y for the young mice, m_a for the aged mice, and unknown standard deviations σ_y for the young mice, σ_a for the aged mice. Here the null hypothesis, H_0, consists of those pairs of normal distributions $(N(m_y, \sigma_y), N(m_a, \sigma_a))$ with $m_y = m_a$, and similarly for the alternative, those with $m_y \neq m_a$.

3. The distribution of the characteristic being measured to detect a change reaction to some treatment can be any distribution pair which has the exact same shape for both young and aged mice.

4. The distribution function of the change in young mice and the distribution function of the change in aged mice are not further constrained beyond the assumption of being continuous and representing independent identically distributed observations for the young mice, and independent identically distributed observations for the aged mice.

Procedures often carried out under the above four assumptions (and which will be introduced and explained later) are respectively:

- the independent two-sample student t test with the *pooled estimate of the standard deviation*

- the Welch modification of the independent two-sample t test

- the Wilcoxon two-sample rank sum test (equivalent to the Mann-Whitney U test)

- the Kolmogorov-Smirnov independent two-sample procedure ∎

Practice Exercises 6.3

1. What would you think is the class of distributions corresponding to Assumption 3. above? Which elements of this class would conventionally be called the null hypothesis, H_0? How would you modify this choice of null hypothesis to make it a more reasonable one?

2. What would you think is the class of distributions corresponding to 4. above? Which elements of this class would conventionally be called the null hypothesis, H_0? How would you modify this choice of null hypothesis to make it a more reasonable one?

In the previous paragraph we alluded to some test procedures which are commonly associated with certain sets of assumptions — as if given sets of assumptions inevitably lead to associated procedures. It is also commonly assumed that if a test procedure was derived under certain assumptions, then use of the procedure is only valid when these assumptions hold. Unfortunately, both these paradigms are counterproductive. Here are some of the reasons:

- It is rare that most assumptions are really satisfied. For instance

  Assumed normal distributions are almost always approximations.

  The assumption that the distributions of two groups have **exactly** the same shape probably holds only rarely.
  The assumption of continuous distribution functions is an idealization made for mathematical convenience — the type of description that simplifies the obtaining of results, but makes little pretense of being a perfect one.

- Even when assumptions appear to be seriously violated, the conclusions resulting from them may turn out to be sufficiently accurate. For example, the student t test often gives valid useful (although not the most efficient) results when applied to uniform distributions.

- Sometimes, simple modifications applied to a test procedure yielding wildly incorrect results, can convert it to a valid efficient procedure (as when the upper and lower 2% of observations are discarded before applying the student t test; this process is called *trimming*).

The important aspect to stress here, is that there is no easy recipe for choosing the test procedures that should be used in various situations. Later, in the discussion of (*discriminating*) *power computations*, the approaches to deciding on appropriate tests will be discussed further.

Logically it would make sense at this stage to develop some theory to guide us in the construction of appropriate tests, and develop criteria for judging the performance of any suggested tests. However, in light of the need to develop some usable skill in the area of testing hypotheses, and the inability to introduce enough theory to make the effort worthwhile in the time frame available, no theory on the construction of tests will be introduced. We will, however discuss the issue of evaluating tests in order to judge whether they are performing in a satisfactory manner. In order to have some examples in mind, we will first introduce a number of the *statistical war horses* of this area. This approach should better prepare us to understand and work with the concept of *power of a test*, when it is introduced.

6.2 Some Commonly Used Statistical Tests

One aspect of testing hypotheses that will become apparent from the examples to follow, is that the **test procedures** being presented **depend critically on both the null hypothesis**, H_0 (associated with the distributions leading to an answer of *no* to the question being investigated) **and the alternative hypothesis**, H_1 (associated with those distributions leading to a *yes*).

Topic 6.2: One-sample Z tests[1]

Suppose we have statistically independent measurements which are normally distributed with unknown mean μ and known standard deviation σ. Such measurements may arise when we are making our measurements with a familiar instrument, and all of the variability comes from this instrument. If our measurements are on some device such as a new energy source (battery) it might be of interest to test whether the total available energy was

1. different from that produced by the old standard
   or (if we knew that the new unit was at least as good as the old standard)
2. greater than the old standard.

The null hypothesis, in either case is that $\mu = \mu_0$, where μ_0 is the average total energy of the old standard.[2]

In Case 1 the alternative hypothesis would be those normal distributions with standard deviation σ and $\mu \neq \mu_0$,

whereas

in Case 2 the alternative hypothesis would be those normal distributions with standard deviation σ) and $\mu > \mu_0$.

If we had available precisely n statistically independent observations $X_1,...,X_n$, Z tests are based on the statistic

$$\bar{X} = \frac{1}{n} \sum_{i=1}^{n} X_i$$

and are of the form:

1. reject the null hypothesis H_0 that $\mu = \mu_0$ in favor of the alternative H_1

   that $\mu \neq \mu_0$ if and only if $\quad |\bar{X} - \mu_0| > c$

---

1. The symbol Z conventionally stands for a standardized normal random variable, i.e., a normal random variable with mean 0 and standard deviation 1. It is used here because all of the Z tests can be viewed as using a statistic which is a standardized normal random variable when the null hypothesis H_0 holds.

2. More precisely, H_0 consists of the normal distributions with mean $\mu = \mu_0$ and standard deviation σ.

(the so-called *two-tailed test*) where c is some constant to be chosen according to how likely you are willing to be wrong when H_0 is true — i.e., how willing you are to be wrong in saying $\mu \neq \mu_0$, when in fact $\mu = \mu_0$.

2. reject the null hypothesis if and only if

$$\overline{X} - \mu_0 > c$$

(one of the one-tailed tests) where c is some constant to be chosen according to how likely you are willing to be wrong when H_0 is true — i.e., how willing you are to be wrong in saying $\mu > \mu_0$, when in fact $\mu = \mu_0$.

and a case we hadn't mentioned,

reject the null hypothesis H_0 that $\mu = \mu_0$ in favor of the alternative H_1 that $\mu < \mu_0$ if and only if

$$\overline{X} - \mu_0 < c$$

(the other *one-tailed* test) where c is some constant to be chosen according to how likely you are willing to be wrong when H_0 is true — i.e., how willing you are to be wrong in saying $\mu < \mu_0$, when in fact $\mu = \mu_0$.

We now compute the constant, c, for the first case (where H_1 is that $\mu \neq \mu_0$), where we want to limit the probability of falsely rejecting H_0 to .01. This condition translates to the equation

$$P_{\mu=\mu_0}\{|\overline{X} - \mu_0| > c\} = 0.01 \qquad\qquad 6.1$$

where the subscript indicates that this probability is to be computed under the condition that $\mu = \mu_0$.

Notice that Equation 6.1 can be written in the form

$$1 - P_{\mu=\mu_0}\{|\overline{X} - \mu_0| \leq c\} = 0.01$$

or

$$P_{\mu=\mu_0}\{|\overline{X} - \mu_0| \leq c\} = 0.99$$

which is equivalent to

$$P_{\mu=\mu_0}\{-c \leq \overline{X} - \mu_0 \leq c\} = 0.99.$$

Using Theorem 4.40 on page 141, we know that $(\overline{X} - \mu_0)/(\sigma/\sqrt{n})$ has a standard normal distribution, so dividing all three terms in the above probability by $\sigma/\sqrt{n}$, we find that the equation which we need to solve for c is

$$P_{\mu=\mu_0}\left\{-\frac{c}{\sigma/\sqrt{n}} \leq \frac{\overline{X} - \mu_0}{\sigma/\sqrt{n}} \leq \frac{c}{\sigma/\sqrt{n}}\right\} = 0.99$$

Now making use of the standard normality of the center term in the above inequality, and the results of Example 2.2 on page 26, we find that $c/(\sigma/\sqrt{n}) = 2.5758$ or

$$c = 2.5758\sigma/\sqrt{n}$$

The quantity c above is called the **critical value** of the test of this H_0 based on the statistic $|\bar{X} - \mu_0|$; the value $\alpha = 0.01$ is called the **significance level** of this test.

To summarize, if we want to test the hypothesis H_0 that $\mu = \mu_0$ against the alternative H_1 that $\mu \neq \mu_0$, with the probability of rejecting H_0 when H_0 is true being chosen as .01 (i.e., a significance level of .01) then we reject H_0 precisely when

$$|\bar{X} - \mu_0| > 2.5758\sigma/\sqrt{n}.$$

Practice Exercises 6.4

1. Determine the critical value, c, for the two other alternatives which go with the null hypothesis H_0 that are given on page 187:

   a. when the probability of rejecting H_0 when H_0 is true being .01

   b. for an arbitrary probability, α, of rejecting H_0 when H_0 is true.

2. Generate a sample of 5 normal random variables with mean 2 and standard deviation 3.

   a. Assuming you know this standard deviation, test the null hypothesis that the mean is 0, versus the alternative that the mean is nonzero, with the probability of rejecting the null hypothesis when it is true set at .05.

   b. Same as part a. with the alternative being that the mean exceeds 0.

   c. Same as part a. with the alternative being that the mean is less than 0.

3. Repeat the previous question where the sample size is 16.

Problem 6.3

It can be shown that when the random variables $X_1, ..., X_n$ are independent, normal with unknown mean μ and (usually) unknown standard deviation σ, the so-called **one sample student t random variable,** t_n given by the formula

$$t_n = \frac{\sqrt{n}(\bar{X}_n - \mu)}{s_n}$$

(where $\bar{X}_n = \frac{1}{n}\sum_{i=1}^{n} X_i$ and $s_n = \sqrt{\frac{1}{n-1}\sum_{i=1}^{n}(X_i - \bar{X}_n)^2}$)

has the *student t distribution with n - 1 degrees of freedom.*
Show how this can be used to test the same hypotheses as the one-sample Z
tests.

These tests are called the **one-sample t tests.**

Topic 6.4: Paired (student) t test

In trying to assess the effect of some treatment (say for high blood pressure) a pair
of measurements may be made on each individual being tested — a measurement
before the treatment and one *following* it. Suppose these measurements are denoted
by (b_i, f_i), $i = 1, 2, ..., n$, and the experimenter wants to see if the treatment had an
effect. We assume that the measurement pairs are statistically independent if they
come from different individuals. However it is reasonable to believe that the
treatment would affect each individual in roughly the same way (lowering blood
pressure of each individual by the same amount, **on the average**). So it is not
assumed that the two elements of any pair are statistically independent. Thus, if we
want to test whether the treatment is effective, the differences, $d_i = f_i - b_i$, are
computed, and one-sample t test of the hypothesis H_0 that $E(d_i) = 0$ is performed
on them.

To see how this procedure behaves, we first generate some data to simulate an
average drop of 17 points (millimeters of mercury) in systolic blood pressure as a
result of the treatment. To account for differences in blood pressure among
individuals with this condition, we assume that systolic blood pressure (SBP) is
normally distributed in this population with mean value 175 mm, and standard
deviation 16 mm. The change in SBP due to treatment will be taken as normally
distributed with mean value -15 mm and standard deviation 5 mm. In the session
window of the Minitab screen, we put the pretreatment measurements in column c1,
the change in SBP in column c4, and the post treatment measurements in column c2,
as shown (or use Calc -> Random Data -> Normal ...).

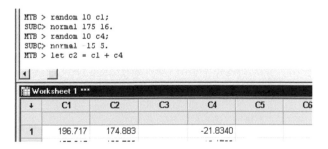

We now go to the menu bar, choosing Stat -> Basic Statistics -> Paired t, to get

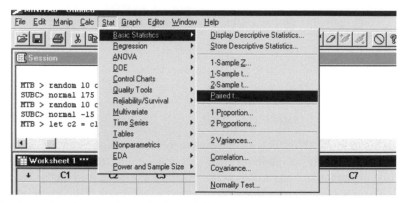

which yields the dialogue box which we fill in as shown.

. c2 = c1 + c4

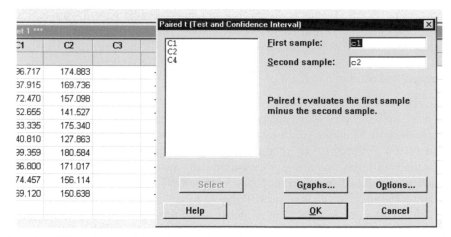

Pressing OK yields the following result on the session screen.

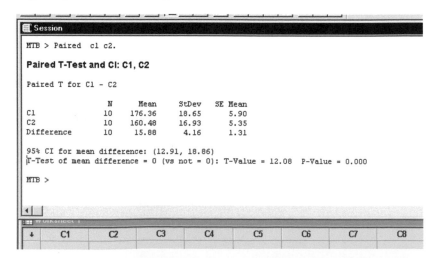

The output indicates that the confidence interval for the $E(d_i)$ did not include 0, so it looks as if $E(d_i)$ is not 0. To clinch this result, we examine the p-value. **Any given p-value indicates that the null hypothesis H_0 would be rejected for any significance level α exceeding this p-value.** So we see that we would surely reject the hypothesis H_0 of *no effect* at any significance level exceeding .0005 (because a p-value indicated as 0.000 means that the p-value is less than .0005). *What was the alternative hypothesis H_1?* It was $E(d_i) \neq 0$ as seen from the phrase *vs not = 0* in the output. The options available from the dialogue box allow for other alternatives.

Practice Exercises 6.5

The next five exercises all refer to Topic 6.4 on page 190.

1. Repeat the computations done in this example, to verify the type of result obtained.

2. Changing only the sample size to 7 find out what happens.

3. Changing only the sample size to 2 find out what happens.

4. Changing only the standard deviation of the elements of column c4 to 10, 15, 30 respectively (still $n = 10$) see how this affects the test.

5. Try some simultaneous changes in the sample size, mean of the differences, and standard deviations of the differences to see their effect.

Topic 6.5: Nonparametric alternative to the one-sample t test

The one-sample **Wilcoxon signed rank test** accomplishes results similar to those of the one-sample t test (see Problem 6.3 on page 189) but is not affected very much if a small percentage of the data is *out in left field*. It accomplishes this behavior by basing its actions on the *ranks* of the $|X_i|$; that is, $|R|_i$ is the position of $|X_i|$ in a list of the $|X_i|$ arranged in algebraic order (assuming no ties in the data).

The Wilcoxon signed rank sum statistic, W_1, may be taken as the sum

$$\sum_{\substack{i \text{ for which} \\ X_i > 0}} |R|_i \quad \equiv S+ \quad ,$$

(the symbol $\equiv$ reads *is abbreviated as*).
The reason for this choice is the following. If the distribution is centered at 0, we would expect this sum to be about half of

$$\sum_{1 \le i \le n} |R|_i \quad = \sum_{i=1}^{n} i \quad = \frac{n(n+1)}{2}$$

i.e.,[1] about $n(n+1)/4$.

---

1. This formula follows easily by letting $S = \sum_{i=1}^{n} i = 1 + 2 + \cdots + (n-1) + n$.

    But also $S = n + (n-1) + \ldots + 2 + 1$.

    Adding these two equations yields $2S = (n+1)+(n+1)+\cdots+(n+1) = n(n+1)$.

If the distribution of the data is shifted to the right, the statistic $S+$ should increase greatly above its null hypothesis expectation, $n(n+1)/4$, since we would expect many more data values to be positive. Similarly if the distribution is shifted to the left, $S+$ should decrease. This leads to the following.

Let the null hypothesis, H_0, be that the median is 0, versus the alternative H_1, that the median is different from 0. The form of the **non-rejection** region for H_0 should consist of those values of $S+$ satisfying

$$c_1 \leq S+ \leq c_2.$$

The **rejection region** for H_0 versus the alternative H_1: the median exceeds 0 should consist of those values of $S+$ satisfying

$$S+ > c,$$

The rejection region for H_0 versus the alternative H_1: the median is less than 0 should consist of those values of $S+$ satisfying

$$S+ < c.$$

This Wilcoxon test is available from the menu bar under Stat -> Nonparametrics. To apply it to matched pairs (of the kind used in the one-sample t test) you have to do the subtractions of the second from the first element of each pair yourself. If the first element of each pair is in c1 and the second in c2, you would put their difference in column c4 via the command let c4 = c1 - c2, and use column c4 in this test. Here are a few exercises to get you familiar with this test, and see how it compares to the paired t test.

Practice Exercises 6.6

1. Generate data of the same type as used in Topic 6.4 on page 190, and compare the results of the paired t test with that of the one-sample Wilcoxon test using the differences of the paired coordinates as its data. (Choose the *testing* version of the Wilcoxon test).

2. Alter the data used in the previous exercise by changing the first coordinate of the first data value by a very large amount (say 500), and see how this changes the results of the previous question.

3. Do some experimentation of your own to see how the paired t test compares with the comparable one-sample Wilcoxon test on the d_i.

Topic 6.6: Independent two-sample Z tests

Often when we want to determine whether a treatment is effective, or compare two treatments, it is impossible to administer both treatments to every individual. For instance, if the treatments are for a disease which gets cured by either treatment, there is no meaningful way to give any patient both treatments — as in the comparison of radical mastectomy with lumpectomy for breast cancer. In research using experimental animals, sometimes the only way to determine the effect of a given treatment is to sacrifice the animal, which tends to make it difficult to give any animal two treatments whose results can be distinguished.

So, suppose we have two treatments, and some measurement of the effectiveness of either treatment on any individual. (The measure might be the time to recovery, or the antibody level at some specified time following the treatment.) It is assumed that the effectiveness of the first treatment is a normally distributed random variable with unknown mean μ_1 and known standard deviation σ_1 and that of the second one is normal with unknown mean μ_2 and known standard deviation σ_2. It is logical to compare the two treatments by using the statistic

$$\frac{\overline{X}_1 - \overline{X}_2}{\sigma_{\overline{X}_1 - \overline{X}_2}}$$

where $\sigma_{\overline{X}_1 - \overline{X}_2}$ is the standard deviation[1] of $\overline{X}_1 - \overline{X}_2$. The **reason for dividing by this standard deviation** is that we want **to compare the difference** $\overline{X}_1 - \overline{X}_2$ (which reflects the value of $\mu_1 - \mu_2$, the average difference in effectiveness of the two treatments) **with** a measure of **how large we would expect** $\overline{X}_1 - \overline{X}_2$ **to be if there was no statistical difference** between the treatments — i.e.,if $\mu_1 - \mu_2$ were 0.

The conventional null and alternative hypotheses are

$$H_0 \text{ that } \mu_1 - \mu_2 = 0$$

and three different possibilities for H_1

$$\text{that } |\mu_1 - \mu_2| \neq 0, \text{ that } \mu_1 - \mu_2 > 0, \text{ and that } \mu_1 - \mu_2 < 0.$$

The corresponding tests for the three alternatives for the 2-sample Z test above, are respectively of the form, to reject H_0

$$\text{when } \left|\frac{\overline{X}_1 - \overline{X}_2}{\sigma_{\overline{X}_1 - \overline{X}_2}}\right| > c, \text{ when } \frac{\overline{X}_1 - \overline{X}_2}{\sigma_{\overline{X}_1 - \overline{X}_2}} > c, \text{ and when } \frac{\overline{X}_1 - \overline{X}_2}{\sigma_{\overline{X}_1 - \overline{X}_2}} < c$$

Choosing the critical value, c, is particularly easy for any of these cases, because from Theorem 4.40 on page 141, the **statistic** (observable random variable)

$$\frac{\overline{X}_1 - \overline{X}_2}{\sigma_{\overline{X}_1 - \overline{X}_2}} \qquad\qquad 6.2$$

has a standard normal distribution when the null hypothesis H_0, that $\mu_1 - \mu_2 = 0$ is true.

---

1. From Theorem 4.40 on page 141 we find that $\sigma_{\overline{X}_1 - \overline{X}_2} = \sqrt{\sigma_1^2/n_1 + \sigma_2^2/n_2}$

   where the meaning of the symbols used should be clear.

Practice Exercises 6.7

1. Suppose there are two routes to work, with expected travel times being unknown. Suppose also, that the standard deviations of the travel times are known values σ_1 = 8 minutes, and σ_2 = 13 minutes. Generate 5 samples of size 6 and size 9 and mean times 35 minutes and 41 minutes, representing observed travel times. Test the null hypothesis that the two routes have the same mean times against the alternative that they are different, each with probability .05 of rejecting the null hypothesis, were it true.

   Unfortunately, Minitab does not supply the two-sample Z test, so you will have to compute the statistic given in expression 6.2 above. You will find it necessary to use Minitab commands in the session screen such as let k1 = mean(c1), let k3 = k2 - k1, let k5 = stdev(c2), let k4 = k3/k2, let k7 = sqrt(k5) etc.

2. Same as previous exercise with the alternative that the second route takes longer on the average than the first one.

The next test we introduce is probably the one most used (and misused) in scientific comparisons.

Topic 6.7: Independent two-sample student t tests

The original independent, pooled variance 2-sample t test was constructed under the assumption of two independent samples $X_1, X_2,...,X_{n_x}$, $Y_1, Y_2,...,Y_{n_y}$ of independent normal random variables, the first with unknown mean μ_x, the second with unknown mean μ_y, where all of the random variables have common, unknown standard deviation, σ. The problem is to test hypotheses concerning $\mu_x - \mu_y$ of the same type as in the two-sample Z tests (presented starting on page 193). It seems reasonable, similarly to what was done in these Z tests, to use as our test statistic, one of the form

$$\frac{\bar{X}_{n_x} - \bar{X}_{n_y}}{\tilde{\sigma}_{\bar{X}_{n_x} - \bar{X}_{n_y}}} \qquad\qquad 6.3$$

where $\tilde{\sigma}_{\bar{X}_{n_x} - \bar{X}_{n_y}}$ is some reasonable estimate of the standard deviation of $\bar{X}_{n_x} - \bar{X}_{n_y}$. In the stated case (equal standard deviations) there is a good estimate of the standard deviation of $\bar{X}_{n_x} - \bar{X}_{n_y}$, namely

$$\tilde{\sigma}_{\bar{X}_{n_x} - \bar{X}_{n_y}} = \sqrt{\frac{(n_x - 1)s_x^2 + (n_y - 1)s_y^2}{n_x + n_y - 2}}, \qquad\qquad 6.4$$

where s_x is the sample standard deviation generated from $X_1, X_2,...,X_{n_x}$ and s_y is the sample standard deviation generated from $Y_1, Y_2,...,Y_{n_y}$ (see Exercise 5.11.1. on page 156). Equation 6.4 gives the *pooled standard deviation* estimate of $X_{n_x} - Y_{n_y}$. Under the given assumptions, when $\mu_x - \mu_y$ = 0, **the independent two-sample t-statistic**

$$t = \frac{\bar{X}_{n_x} - \bar{X}_{n_y}}{\sqrt{\dfrac{(n_x - 1)s_x^2 + (n_y - 1)s_y^2}{n_x + n_y - 2}}} \qquad\qquad 6.5$$

has the student t distribution with $n_x + n_y - 2$ degrees of freedom. It can be used to test hypotheses just as we did with the independent two-sample Z tests (page 193).

Practice Exercises 6.8

To do these exercises, use the Minitab two-sample t test — select from the menu bar
Stat -> Basic Statistics -> 2-sample t (choose *assume equal variances*)

1. Suppose there are two routes to work, with expected travel times being unknown. Suppose also, that the standard deviations of the travel times are known values $\sigma_1 = 8$ minutes, and $\sigma_2 = 8$ minutes. Generate 5 samples of size 6 and size 9 and mean times 35 minutes and 41 minutes, representing observed travel times. Test the null hypothesis that the two routes have the same mean times against the alternative that they are different, each with probability .05 of rejecting the null hypothesis, were it true. Preserve this data set for future use.

2. Same as previous exercise with the alternative that the second route takes longer on the average than the first one.

Unfortunately, the equal variance option does not always give trustworthy answers if σ_x/σ_y is far from the number 1. In practical situations you should check that σ_x/σ_y is close to 1, justifying the use of the equal variance option. To do this, you might want to obtain confidence intervals for the ratios σ_x/σ_y, (see Exercise 5.11.2. on page 157).

It is always **safer to use the Welch modified two-sample t test** (the one which avoids choosing the equal variance option). In this case the test statistic used has a more trustworthy estimate of $\sigma_{\bar{X}_{n_x} - \bar{X}_{n_y}}$, namely $\sqrt{s_x^2/n_x + s_y^2/n_y}$ suggested by Footnote 1. on page 194, and makes a better choice for computing the degrees of freedom,[1] dependent also on how different s_x and s_y are. The confidence interval that is provided also helps determine if there is a practical significance to an observed difference.

The confidence interval provided by both of the independent two-sample t tests is based on the facts that

---

1. This degrees of freedom computation is not as firmly based theoretically as we might like, and is somewhat messier to do by hand; but it is quite trustworthy in general, with the messier details handled by Minitab (and most other statistical packages as well).

- the unobservable random variable

$$\frac{(\bar{X}_{n_x} - \bar{X}_{n_y}) - (\mu_x - \mu_y)}{\sqrt{\dfrac{(n_x - 1)s_x^2 + (n_y - 1)s_y^2}{n_x + n_y - 2}}} \tag{6.6}$$

when the variances are equal (and the pooled variance estimate is valid) has a student t distribution with $n_x + n_y - 2$ degrees of freedom.

- in the Welch modified case, the unobservable random variable

$$\frac{\bar{X}_{n_x} - \bar{X}_{n_y} - (\mu_x - \mu_y)}{\sqrt{\dfrac{s_x^2}{n_x} + \dfrac{s_y^2}{n_y}}} \tag{6.7}$$

has approximately a student t distribution with n_{Welch} degrees of freedom (where n_{Welch} is the approximate degrees of freedom computed by Welch).

Practice Exercises 6.9

1. The following is real data arising from aged (a) and young (y) mice. The values represent measurements on untreated aged mice (ctr), aged mice at 2 hours and 9 hours after treatment, then young untreated mice, and these mice 2 hours and 9 hours after treatment. Every number represents a different mouse. The questions of actual interest were:

   a. Did the young mice in each of the columns differ from the older mice in the corresponding columns?

   b. Did the treatment on aged mice have an effect after 2 hours and after 9 hours, and the same question for the young mice. Answer these seven questions. In particular, if you failed to find a difference, does it appear that this might be because there was not enough data (assuming that the biggest 95% guaranteed difference you found was of importance)? After you have read this data set in, preserve it for future use.

a2ctr	a2t	a9t	y2ctrl	y2t	y9t
165527	211430	457610	9337.40	7115.9	6810.1
132526	214373	341221	6473.40	6727.2	11095.6
140065	205718	740299	5388.10	6450.8	18553.6
	399989	616518	5090.63	7630.9	27435.9
		372390	4916.40	8780.9	42517.0
		397412	4442.99	10086.5	57146.2

2. Suppose there are two routes to work, with expected travel times being unknown. Suppose also, that the actual standard deviations of the travel times are $\sigma_1 = 2$ minutes, and $\sigma_2 = 6$ minutes. Generate four pairs of samples of size (10,40), (40,10), respectively, with mean times 35 minutes and 50 minutes, representing observed travel times (and preserve the data sets generated, for future use).

   a. Test the null hypothesis that the two routes have the same mean times against the alternative that they are different, each with probability .05 of rejecting the null hypothesis, were it true, by both the pooled variance, and independent two-sample.

   b. Examine the confidence intervals associated with these tests, to determine how the sample sizes determine how much differently the pooled and unpooled variance procedures behave.

Problem 6.8

If F_X is normal and F_Y is some other distribution function, will the two sample t test based on enough data have a high probability of rejecting the null hypothesis that $F_X = F_Y$?

Topic 6.9: The independent two-sample Wilcoxon test (aka the Mann-Whitney test)

In Exercise 6.6.2. on page 193 you found (I hope) that the one-sample t test can be greatly influenced by a single *outlying* observation. Such an observation tends to prevent the t test from rejecting the null hypothesis. As you might expect, t tests do not tend to do that well on distributions which are highly unsymmetric. For this reason the two-sample Wilcoxon test less susceptible to such disruptions[1] was developed. This test is referred to as *nonparametric*, but this term really means that the test works reasonably well on a wider class of distributions than the class on which the original parametric test (the two-sample t test) works well.

The Wilcoxon two-sample rank sum test is constructed from two independent samples $X_1, X_2,...,X_{n_x}$ and $Y_1, Y_2,...,Y_{n_y}$ each of independent identically distributed observations, where we list the smaller of the two samples first (i.e., $n_x \leq n_y$). Let F_X and F_Y be the distribution functions of the X_i and Y_j. The test will nominally be of the hypothesis H_0 that $F_X = F_Y$ versus one of three possible alternatives H_1 that F_Y is of the form given respectively by

$$F_Y(y) = F_X(y-c) \text{ for all } y, \text{ for some } c \neq 0 \qquad\qquad 6.8$$

$$F_Y(y) = F_X(y-c) \text{ for all } y, \text{ for some } c > 0 \qquad\qquad 6.9$$

$$F_Y(y) = F_X(y-c) \text{ for all } y, \text{ for some } c < 0 \qquad\qquad 6.10$$

---

1. Being insensitive to *outliers* may not always be desirable. So we should always judge carefully whether a test is doing what it is supposed to.

The test procedure is the following. Combine the two samples into one sample, $X_1, X_2,...,X_{n_x}, Y_1, Y_2,...,Y_{n_y}$. Form a new sample by replacing each of the elements of this last sequence by its rank in this sequence. (Recall the definition of *rank*, at the beginning of the section on nonparametric alternative to the one-sample *t* test on page 192.) This replacement sequence is $R_1, R_2,...,R_{n_x}, R_{n_x+1}, R_{n_x+2},...,R_{n_x+n_y}$, where $R_1, R_2,...,R_{n_x}$ are the ranks of the X_i.

Recall the formula $1 + 2 + 3 + ... + n = n(n+1)/2$ (see Footnote 1 on page 192). From this it follows that the average rank of the combined sequence is $(n_x + n_y + 1)/2$. Therefore under the null hypothesis H_0 that $F_X = F_Y$, the expected value

$$E\left(\sum_{i=1}^{n_x} R_i\right) = n_x(n_x + n_y + 1)/2$$

If F_X is just the same as F_Y, but shifted to the right, we would expect $\sum_{i=1}^{n_x} R_i \equiv Sxr$ to increase, while if it was F_Y shifted to the left, we would expect this sum to decrease. The Wilcoxon independent two-sample test is therefore

The form of the **non-rejection** region for H_0 corresponding to alternative 6.8 is

$$c_1 \le Srx \le c_2$$

The form of the **rejection region** for H_0 versus the alternative 6.9 is $Srx > c$,

The form of the **rejection region** for H_0 versus the alternative 6.10 is $Srx < c$,

All of this may be accomplished by first putting your samples in two columns, say c1 and c2, and then choosing from the menu bar
        Stat -> Nonparametrics -> Mann-Whitney which gets the following

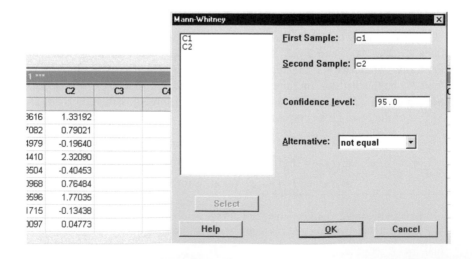

Next fill in this dialogue box appropriately, our choices appearing there now, and click OK, to get a result similar to the one shown below.

Mann-Whitney Test and CI: C1, C2

```
C1        N =  10     Median =     0.090
C2        N =  20     Median =    -0.080
Point estimate for ETA1-ETA2 is      0.161
95.5 Percent CI for ETA1-ETA2 is (-0.482,0.788)
W = 172.0
Test of ETA1 = ETA2   vs  ETA1 not = ETA2 is significant at 0.4679
```

	C1	C2	C3	C4	C5	C6	C7
1	0.83616	1.33192					
2	0.07082	0.79021					

Practice Exercises 6.10

Modify the questions in Practice Exercises 6.8 on page 196 and Practice Exercises 6.9 on page 197 so that you can apply the 2-sample Wilcoxon test to the appropriate data sets you preserved . Compare these results with those from pooled variance and with those from the unpooled variance two-sample *t* tests.

Problem 6.10

If F_X and F_Y are different distribution functions, will the two-sample Wilcoxon rank sum test based on enough data have a high probability of rejecting the null hypothesis that $F_X = F_Y$?

Topic 6.11: The chi-square tests of homogeneity and independence

In trying to assess the effects of exposure to various pollutants we often have reason to compare the distributions of disease severity in several populations. For instance, we might have the following categories of severity of lung dysfunction: absent, mild, medium, severe, life threatening. The populations we might want to compare could be : near factory, suburbs and rural.

We might take samples of preassigned size from each of these populations, and enter the number of people in each category for every population in a *contingency table* of the type shown in Table 6.1.

Table 6.1

Category	absent	mild	moderate	severe	life threatening
Population					
near factory	5	7	15	9	7
urban	12	17	15	6	5
rural	29	16	14	8	6

The information we are seeking is whether the populations are statistically the same, insofar as their distribution into categories (the alternative hypothesis is that the distributions are not all the same); in the illustration, whether the severity of symptoms is distributed in the same way for all three populations. We form the so-called **chi-square statistic**, often written symbolically as

$$\chi^2 = \sum_{\text{all cells}} \frac{(\text{Obs} - \text{Exp})^2}{\text{Exp}} \qquad 6.11$$

In the general case the table has R rows and C columns, and there are $R \times C$ terms in this sum, one for each entry in the contingency table. In the row r, column c entry in the sum

Obs stands for the value in this cell (e.g., the row 2 col 3 entry is 15 in the table above). The observation, Obs, in the r-th row, c-th column will be denoted by $N_{r,c}$.

The quantity **Exp** represents our estimate of the expectation of the cell entry, $N_{r,c}$, if the null hypothesis H_0 that all populations are statistically the same (with regard to category distribution) is true.

Now if H_0 is true, then the reasonable estimate of the probability of **category c** in all of these populations is just

$$\hat{p}_c = \frac{\text{total \# of category c observations in the table}}{\text{total \# of observations in the table}} = \frac{\displaystyle\sum_{\text{all } r} N_{r,c}}{\displaystyle\sum_{\text{all } r,c} N_{r,c}}$$

Hence, our best estimate of what to expect in cell r,c when H_0 is true, Exp, would be given by $N_{r.} \times \hat{p}_c$, where $N_{r.} = \displaystyle\sum_{\text{all } c} N_{r,c}$ is the size of population r.

So the Chi-Square statistic is

$$\sum_{all~~r,\,c}\left[\left(N_{r,\,c}-N_{r.}\frac{\displaystyle\sum_{all~~r'}N_{r',\,c}}{\displaystyle\sum_{all~~r',\,c}N_{r',\,c}}\right)^{2}\middle/\left(N_{r.}\frac{\displaystyle\sum_{all~~r'}N_{r',\,c}}{\displaystyle\sum_{all~~r',\,c}N_{r',\,c}}\right)\right]$$

Recall that the sum of the observed $N_{r,\,c}$ over c is $N_{r.}$; let $N_{.c}$ denote the sum of the observed $N_{r,\,c}$ over r, and $N_{..}$ denote the sum of $N_{r,\,c}$ over all r,c. Then we may write the Chi-Square statistic in the form

$$\chi^2 = \sum_{all~~r,\,c} \frac{(N_{r,\,c}-N_{r.}N_{.c}/N_{..})^2}{N_{r.}N_{.c}/N_{..}}\qquad\qquad 6.12$$

If the null hypothesis H_0 that all populations are statistically the same (with regard to category distribution) is true, we would expect χ^2 to be much smaller than if this hypothesis were seriously violated. In fact, using Central Limit Theorem arguments (Theorem 4.38 on page 135), together with the fact that a chi-square variable with n degrees of freedom is the sum of squares of n independent standard normal variables, shows that under H_0, χ^2 has approximately a chi-square distribution with $(R - 1)(C - 1)$ degrees of freedom.

Your contingency table should first be entered in the worksheet in contiguous columns, each with the same number of integer entries (corresponding to the values $N_{r,\,c}$).   Once you have a contingency table, the chi-square test can be run by choosing as follows from the menu bar: Stat -> Tables -> Chi-square Test

You will see the following dialogue box

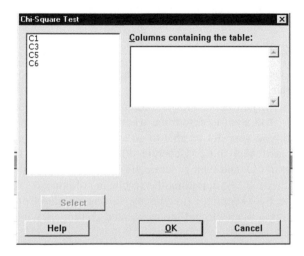

Indicate the columns your table is in, and then press OK. I entered the following numbers in columns c9-c12.

9	C10	C11	C12	C1:
8	7	5	12	
6	13	11	13	
19	23	17	14	

After pressing OK, the following results appeared on the session screen.

```
 Session

              e        C10       C11       C12     Total
    1         8          7         5        12        32
            7.14       9.30      7.14      8.43

    2         6         13        11        13        43
            9.59      12.49      9.59     11.33

    3        19         23        17        14        73
           16.28      21.21     16.28     19.24

 Total       33         43        33        39       148

 Chi-Sq =  0.105 +  0.568 +  0.639 +  1.509 +
           1.343 +  0.021 +  0.208 +  0.246 +
           0.456 +  0.151 +  0.032 +  1.425 = 6.702
 DF = 6, P-Value = 0.349
```

The decimal numbers appearing are the estimated null hypothesis expectations.

The chi-square test for independence applies to discrete integer valued random variables taking on the same finite set of values. The categories in the test of homogeneity are replaced by the values of the random variables; the populations by the symbols for the random variables themselves.

Folk lore has it that each cell in the contingency table should have a value in it no smaller than 5.

Problems 6.12

1. Suppose you went to Las Vegas, and spent some time at the craps table. Further, suppose you kept track of the dice toss results, where they used the same pair of dice for 200 tosses, and then switched dice for the next 200. You want to determine whether or not the dice loadings changed. Well, you may not be able to go to Las Vegas just now, but you can do the next best thing. Generate 400 fair die tosses (put the results in column c1) and then generate 400 die tosses where the probabilities are .1, .2, .3, .1, .1, .2, putting the results in column c2. You will need to do some bookkeeping (maybe by sorting columns c1 and c2, maybe writing a macro to get the associated contingency table). Run the chi-square test.

2.   Do a few more similar runs.

3.   How could you use the ideas developed here to test whether a given pair of dice was a fair one?

4.   Show that the chi-square test of independence represents a reasonable test of the hypothesis of independence of the random variables involved.

Topic 6.13: Other tests

There are many other tests of hypotheses developed to have specialized properties. **The Kolmogorov-Smirnov (K-S) two sample tests** are based on the empirical cumulative distribution functions, and will, with sufficient data, have a very high probability of **distinguishing any two distributions which are, in fact, different** (in contrast to the two-sample t test, and, as we will see, to the Wilcoxon two-sample test). The test statistic for the K-S is the maximum vertical distance between the two empirical distribution functions. There is also a K-S test of fit, which uses as its test statistic the maximum vertical distance between the observed empirical distribution function, and the hypothesized distribution function

The **Fisher exact test** has the same form as the chi-square test, but is allegedly more reliable — at the cost of making the computer work much harder. There are tests of statistical independence, tests of inequality of variances and many more.

There are some tests which have little merit, and they are often used because they sometimes provide the answers people want, as opposed to reflecting the true situation. In the next section we will see how simulation can often let you know how well any test is doing in a variety of different situations.

Topic 6.14: *P*-values

In the tests that we have examined we chose the critical value to control the probability of rejecting the null hypothesis, H_0, when H_0 was true — i.e., when the governing distribution was one of the null hypothesis distributions. In many cases the maximum probability of rejecting H_0 when H_0 is true, is the probability of falsely rejecting H_0 for all distributions in the null hypothesis. To be more general, we give the following definition.

Definition 6.15: Significance level of a test, *p*-value of test statistic

The smallest number not exceeded by any probability of falsely rejecting H_0 is called the *significance level* of the test being considered, and is usually denoted by the Greek letter α.

The **p-value associated with a test which has been performed is the smallest significance level at which H_0 will be rejected for the given value of the test statistic being used.**

□

So the *p-value* is associated only with null hypothesis distributions and the observed value of the test statistic. It therefore tells little about the probability of rejecting the null hypothesis when some given alternative governs the data. The main use of the p-value is to let you know whether you have or have not rejected the null hypothesis for a specified significance level. If you were told that the p-value was .001, you know that if you chose a significance level (which, for practical purposes you can think of as the maximum probability of falsely rejecting H_0) of α = .01, then you would have rejected H_0 with the data you had. This explains why p-values are almost always presented with tests.

Topic 6.16: Setting up tests of hypotheses

The basic steps in formulating tests of hypotheses to answer real problems may be summarized as follows.

- **Phrase a *yes/no* question about the system being studied**:

  For example
  > *Does treatment 1 have the same (average) effect as treatment 2?*
  versus
  > *Is treatment 1 better than treatment 2?*

- **Collect data to help answer the posed question.**

  For example, collect data $X_1, X_2,...,X_{n_x}$ on treatment 1 recovery times and $Y_1, Y_2,...,Y_{n_y}$ on treatment 2 recovery times.

- **Decide on the form of the probability distribution of the collected data.**

  Here maybe you assume that each of the samples consists of independent identically distributed normal random variables, with unknown means and standard deviations (μ_x, σ_x) and (μ_y, σ_y) respectively.

- **Translate the original yes/no question into a quantitative one concerning hypotheses related to the chosen sets of probability distributions H_0, H_1.**
  In the example being considered, the hypothesis H_0 might be the subset of all the chosen distributions with $\mu_x = \mu_y$, while H_1 would then consist of the subset with $\mu_y < \mu_x$ (Note that in deciding on the form of the possible probability distributions, we should have omitted those with $\mu_x < \mu_y$.)

- **The next step is to choose a test statistic on which to base the test.**

  Here it seems reasonable to use the modified unpooled form of the t statistic

  $$\frac{\bar{X}_{n_x} - \bar{Y}_{n_y} - (\mu_x - \mu_y)}{\sqrt{s_x^2/n_x + s_y^2/n_y}}$$

  developed by Welch.

- **Finally, the way to use the test statistic for testing the hypothesis must be decided on.**

The form of the test may be an obvious one, but details, such as the choice of sample size may need to be determined by specific properties that the test should have (such as adequate discriminating power, a subject to be discussed in the next section). In the illustration being used, the form of the test seems like it should be to reject H_0 if $t_n > c$, where the critical value, c is some suitably chosen constant.

The above is a simplified outline of the steps that seem reasonable in setting up tests of hypotheses. Each of these steps may require careful, time-consuming consideration, especially if the problem is not a standard one.

6.3 Types I and II Errors and (Discriminating) Power

Definition 6.17: Power function, type I and type II errors

A function whose input is some probability distribution, and whose output is the associated probability that a particular hypothesis test will reject the null hypothesis H_0 when this distribution governs the data, is called a **power function** for this test. A **type I error** consists of rejecting H_0 when the true distribution lies in H_0; a **type II error** is failure to reject H_0 when the true distribution lies in H_1. So the power function gives the probability of type I error for those distributions in H_0, and it gives 1 - P{type II error} for those distributions in H_1.

❑

If you know its power function for all possible distributions governing the data, then you know how the test behaves. Usually there will be some probability distributions for which you strongly hope that the power function will be small, and there will be some probability distributions for which you strongly hope that it will be large. Often there will be some to which you are indifferent. For instance, a polling company would like to be reasonably sure that its prediction of the winner is valid when the margin of victory/defeat is outside the range (49%; 51%); not that they don't prefer to be right in this region — but they know they are not likely to be blamed for missing such a close call.

In testing hypotheses, in most cases there is some set of possible distributions of the data for which it is desired to compute the power function. Knowing the test, in theory it is possible to compute the power function at any given probability distribution which could describe the data. But that is easier said than done, since the entire subject of mathematical probability deals with such types of computation (and we know that there are, and always will be, many such problems unsolved).

For some of the more popular tests the work of computing the power for certain probability distributions that could govern the data has been done, and the results made available in many computer packages, including Minitab.

Example 6.11: Power of the independent two-sample t test

Suppose we would like to find the discriminating power of an independent two-sample t test where the difference in means is 4, under the restriction $n_x = n_y$, with two different choices for n_x, 15, and 19. Going to the menu bar, choose
    Stat -> Power and Sample Size   to obtain what you see below.

Now choose 2-Sample t to reach the following.

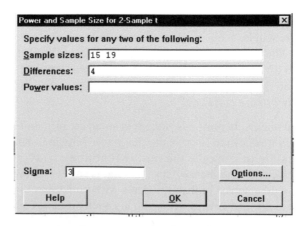

Fill it in as shown in the first two areas, and fill in the chosen standard deviation, (Sigma),select OK, and the result will appear as shown, in the session screen.

```
MTB > Power;
SUBC>   TTwo;
SUBC>       Sample 15 19;
SUBC>       Difference 4;
SUBC>       Sigma 3.
```

Power and Sample Size

```
2-Sample t Test

Testing mean 1 = mean 2 (versus not =)
Calculating power for mean 1 = mean 2 + difference
Alpha = 0.05  Sigma = 3

              Sample
Difference     Size    Power
         4       15   0.9411
         4       19   0.9792
```

The output is self explanatory. If you felt that the power is more than you need, you could reduce the sample sizes, or the difference, and try again.   ■

In the illustration just presented, the power depended on the unknown standard deviation σ, but a choice of σ had to be made anyway. This choice is usually based on past history, but is often made in the hope that agencies that make grants will believe the choice valid. The *difference* in this example is a hypothetical difference between mean values, and is usually chosen to yield reasonably high power, and (it is hoped) will convince the granting agencies that being able to distinguish such differences is a worthwhile achievement.

It should be noted that this program can be used to determine how many observations are needed to achieve a given power at a specified difference in means (for the chosen standard deviation) and what difference in means will yield a given power with given sample sizes.

Practice Exercises 6.12

Try some of the variations of the power/sample size program, including different specified values of 2 of the 3 choices, and different distributions.

The biggest flaw in these **theoretical power computations** is that in a great many cases, the actual distributions governing the data are far from the postulated ones, and are ones not amenable to theoretical computation. So the results produced make fine fiction, but are somewhat lacking in the *reality* department. There is, however, an alternative approach which largely overcomes these flaws, and it is outlined in the following section.

6.4 The Simulation Approach to Estimating *Power*

In this section we will outline an approach to estimation of discriminating power, which allows more flexibility and reality than the theoretical approach introduced above. The quality of the results is dependent on both the quality of the random number generator and the sample size being used. We illustrate with the Wilcoxon two-sample test (Topic 6.9 on page 198).

First, decide on the distributions from which the two samples are to be drawn. Just about any discrete distribution which yields only a finite number of possible values, or the available ones under Calc -> Random Data... can be used. For this example, the x and y samples are of size 13 and 17 respectively, both coming from Cauchy distribution with location parameter 0.0 and scale parameter 3.0 (the distributions being the same to simulate the null hypothesis holding). Next, choose a critical value to yield a significance level (Definition 6.15 on page 204) α (say .05).

Now choose the test statistic, e.g., the sum of the x ranks, which would be associated with the Wilcoxon independent two-sample test (starting page 198). Some way of computing this test statistic is needed. At worst, this can be done by running the test itself, but then it is necessary to disentangle the desired information from the test's output; since this simulation may be running this hundreds of times or more, this is a crude approach (although it is possible to write a program to untangle things). In the macro to be presented, here is the approach taken to handle the generation of simulated data, computation of the Wilcoxon test statistic, its conversion to the *number of rejections on the given trial* (0 or 1), and storage of these values in some Minitab column.

First, the menu bar choice

Calc -> Random Data -> Cauchy

was used to choose a sample, in order to learn the line commands to be used in the macro. Running the above yielded the following dialogue box.

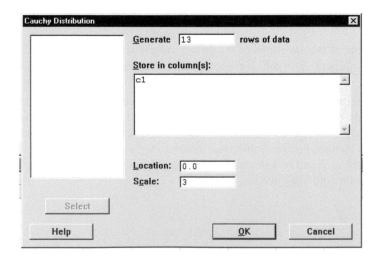

Selecting OK yielded the following on the session screen.

```
MTB > Random 13 cl;
SUBC>   Cauchy 0.0 3.
MTB >
```

	C1	C2	C3	C4
1	6.3584			
2	0.9196			

Worksheet 1 ***

Since it can be shown that for sample sizes bigger than 9 or 10, the test statistic, Sxr (sum of the x ranks) is approximately normal, with mean $n_x(n_x + n_y + 1)/4$ and standard deviation $\sqrt{n_x n_y(n_x + n_y + 1)/12}$, to get the first guess for the correct critical values for the two-sided test (for the alternative H_1, that $F_X \neq F_Y$) will be found by first noting that approximately

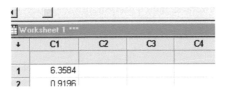

$$P\left\{\left|\frac{Sxr - n_x(n_x + n_y + 1)/2}{\sqrt{n_x n_y(n_x + n_y + 1)/12}}\right| > z_{1-\alpha/2}\right\} = \alpha \quad \text{i.e.,} \quad P\left\{\left|\frac{Sxr - n_x(n_x + n_y + 1)/2}{\sqrt{n_x n_y(n_x + n_y + 1)/12}}\right| \leq z_{1-\alpha/2}\right\} = 1 - \alpha$$

With the same kind of manipulations done in following Equation 6.1 on page 188 the nonrejection region for the rank sum statistic, Sxr, is found to be those Sxr in the interval

$$n_x(n_x + n_y + 1)/2 \pm z_{1-\alpha/2}\sqrt{n_x n_y(n_x + n_y + 1)/12}$$

(where $z_{1-\alpha/2}$ is the upper $1 - \alpha/2$ point on the standard normal distribution). If $\alpha = 0.05$, then $z_{1-\alpha/2} = 1.975$, and first guess for the critical values, (c_-, c_+) is $(154.67, 248.33)$.

To compute the Wilcoxon rank sum statistic for any one trial, the columns in which the samples were put are stacked in some other column. Then using the *rank* function on this stacked column the ranks are calculated. Next the sum of the first n_x ranks are computed, and put into some other column in the row of this other column equal to the trial number. Then this last element is converted to 0 (nonrejection) or 1 (rejection) depending on its value, and put into the same row in yet another column.

Outside the macro, using the *let* command, we put our choices for certain important constants in so-called *constant variable locations*, k1 for n_x, k2 for n_y, k3 for the number of simulations, k8 for c_-, k9 for c_+, k10 for the y location parameter and k11 for the y scale parameter.   Here is the macro, *iwilpow.mac* (the prefix i being put on because this might be the inner loop of some front-end macro).

```
GMACRO
iwilpow  #template (required)


erase c1-c7  # to get a clean screen
do k4 = 1:k3  # k3 is number of simulations  k4 is  index of this loop
 random k1 c1;  # generates data satisfying null hypothesis  k1 is nx
 cauchy 0 3.     # could make these variable constants
 random k2 c2;  # generates data satisfying alt hypothesis   k2 is ny
 cauchy k10 k11.


stack c1 c2 c4  # so that column c4 can be ranked


let c5 = rank(c4) # actually assigning ranks stored in c5


let k5 = 0  # initialization to allow correct computation of test statistic
 do k6 =1:k1  # This loop forms the test statistic, assigned to  k5
          # k6 is index of this loop
 let k5 = k5 + c5(k6)
 enddo


 let  c6(k4) = k5 # c6 has the rank sum statistic for trial k4  in row k4


# k8 and k9 are our critical values,  (k8,k9) is nonrejection region
if((k5 >= k8)  &  (k5 <= k9))
  let  c7(k4) = 0 # puts 0 to row k4 of c7 if Ho is not rejected on trial k4
else
  let  c7(k4) = 1 # puts 1 to row k4 of c7 if Ho is rejected on trial k4
   #  c7 has the number of rejections on each trial (0 or 1)
endif
 enddo
ENDMACRO
```

The command $\%c$:\iwilpow was run, with $n_x = k1 = 13, n_y = k2 = 17$, #simulations = $k3 = 1000, yloc = k10 = 0, yscale = 3$. (It runs about 300 simulations per minute.) Column c7 was chosen in the macro iwilpow.mac for listing the results of each trial (1 for rejection, 0 for non-rejection). When it concluded and mean c7 was entered on the session window, we obtained .054. Since this indicates a proportion of rejections near the nominal .05, it seems as if the normal approximation is a good one. By the methods developed in Theorem 5.13 on page 166, using the macro clmsrch (in Example 5.8 on page 166) the .95 level confidence interval for the probability of rejection is (.0408,.0699), so this gives us even more faith in the normal approximation used to make a first guess at the critical values.

Practice Exercises 6.13

1. Run the macro *iwilpow.mac* with the parameters just listed, and with all of these parameters the same except for the y scale parameter (k11) which should be changed from 3 to 13. What does your result appear to tell you about the Wilcoxon test's ability to distinguish between distribution functions which can be very different from each other?

2. Find out how well the Wilcoxon independent two-sample test distinguishes between a normal distribution with mean 0 and standard deviation 1, and an exponential distribution with mean 3 using samples of sizes 13 and 17 respectively for the normal and exponential data.

Problems 6.18

1. Modify the macro *iwilpow.mac* to estimate the power at any chosen normal distribution of the pooled variance independent two-sample t test with sample sizes 13 and 17, using this test's usual critical values.
   Hints: The critical value can be chosen from the t distribution with 13 degrees of freedom tables. You obviously must replace the Cauchy distributions with normal ones, and you must replace the rank statistic computation with the t statistic computation. Run this modification, with the x distribution being normal, with a sample size of 13, and having a mean of 0, and a standard deviation of 3, and with the y distribution being normal with a sample size of 17, a mean of 1.5, and a standard deviation of 3 to compare with the results you get from a theoretical power analysis.

2. Modify the macro *iwilpow.mac* to mimic the Welsh modified version of the independent two-sample t test at the .05 significance level. Run a few simulations, preserving the data, to see if you get the same results by applying the Welch unpooled variance test to this data. Use the results of this and the previous problem to compare the Welch and unpooled variance versions when the standard deviations are not the same.
   Hints: In doing this modification, you need to use as the denominator, the expression $\sqrt{s_x^2/n_x + s_y^2/n_y}$, but you must determine the critical value by simulation (since it is not usually easy to determine except by going to the statistical journal literature).

6.5 Some Final Issues and Comments

A great deal of effort is spent by many people learning statistics to find some way of committing to memory answers to the questions of

- whether to use one-tailed or two-tailed tests,
- whether to use a paired or unpaired test,
- whether to use parametric or nonparametric tests.

The question of whether to use one- or two-tailed tests should only be decided in favor of one of the one-tailed tests if there is virtually no possibility that the alternative could be one of those associated with the tail that would be neglected. (For instance, suppose you were trying to sell your house, and you had a rock solid offer from the bank to buy it at some specified value. Then you could try to test the hypothesis that you can sell your house at the specified value, versus the alternative of being able to sell it at a higher value.) This is a rather rare situation, so that the two-tailed tests (and the associated confidence intervals) are usually recommended. By contrast, sometimes one-sided confidence limits are useful, for example, if you only need an upper bound for a standard deviation.

Whether a paired or an unpaired test should be used, like the previous question, is dependent on the assumptions you make — i.e., on the class of models that are being used. If the realities of the situation under which the data was produced provide convincing evidence that there is less variability in the difference between two elements of the same pair than in the difference between two arbitrary elements, then the paired test is more appropriate. Otherwise the independent sample test is the one to use. For instance, suppose we were comparing lifetimes of pairs of airplane engines maintained under one of two procedures. If we just chose the engines arbitrarily, then the two-sample procedure would seem appropriate. However, if each pair came from the same airplane, and thus both engines were subjected to the same external stresses, a paired test seems reasonable.

In trying to decide on a parametric as opposed to a nonparametric test, the real question is whether the departure from the assumptions on which the parametric test was based are of the type that will be likely to alter the test behavior seriously. So if large outliers are likely to occur, the t test seems inadvisable.

All of these are issues about the type of model that is appropriate to solving the problem being addressed. To use statistical procedures intelligently, it is important to pay attention to the assumptions suited to the problem being investigated. Applying an inappropriate procedure cannot be compensated for by computer expertise or a large repertoire of statistical methods, many of which are not well understood.

In setting up a class of models on which an investigation is to be based, it is important to include a wide enough class of distributions so that it is likely that a sufficiently good description in this class to answer the questions of interest. In the testing hypothesis area, this means that the null hypothesis class and the alternative hypothesis class between them include such a description. Assuming normality when it is substantially violated can produce misleading conclusions, but failing to use the normality assumption when it fits well can lead to failure to draw reasonable conclusions.

At the beginning of this chapter it was mentioned that the first phase of many scientific investigations often consists of hypothesis tests — to see if there is any effect at all. Unfortunately it is the policy of many granting agencies and journals to insist that all problems be phrased as tests of hypotheses, in what is referred to as *hypothesis-driven research*. Many statisticians feel that it is often preferable to phrase such problems as attempts to estimate the size of the difference between two treatments or processes. The confidence intervals that would result from phrasing problems in this way not only answer the question of whether there is a difference, but quantify the size of this difference, or indicate that no difference was detected because of an inadequate amount of data, or an inordinate amount of variability in the data. We stress that when they are available, confidence intervals and confidence limits can usually be used to test hypotheses. For example, a confidence interval for the difference $\mu_x - \mu_y$ from two independent normal samples can be used to make the judgment as to whether or not $\mu_x = \mu_y$, ordinarily done by the two-sample t test.

Outliers (values whose distance from most of the data is large) are a constant source of difficulty for investigators. What to do about them depends on how the results of the investigation are to be used. Several situations come to mind.

For instance, if outliers cause a treatment to fail on some percentage of patients, or a process to fail on some percentage of manufactured units, they can legitimately be tossed out if

- they are impossible physiologically or physically
- caused by an evident error, which is, however, not correctable
- it is made clear, and is acceptable, that the inference to be drawn is to the subpopulation of non-outliers.
  Such might be the case if a treatment for symptoms of a non-threatening disease failed on a certain proportion of patients, but otherwise did no harm.
  It would certainly not be allowable in situations where the outliers caused an unacceptable number of fatalities (a number which would be different for treatment of dangerous diseases, versus treatment of non-dangerous ones).

A completely unacceptable excuse for tossing out outliers is that the investigator was displeased with the results.

Finally, it should be kept in mind that models are only descriptions, and should not be stretched beyond sensible limits — certainly not beyond the range of the observed data, unless there is firm justification for it, justification arising from a thorough understanding of the basics of the underlying mechanism.

Review Exercises 6.14

1. If a new treatment is to be compared with the old standard, what determines whether you should use a one- or a two-tailed test?

2. What determines whether a paired test or an unpaired test should be used for determining if there is a change in the mean value?

3. Referring to Definition 6.1 on page 182, what is the sample space for the problem of Example 6.1 on page 182?

4. Phrase the one-sample Z test (Topic 6.2 on page 187) in the terms put forth in Definition 6.1 on page 182.

5. Same as previous question for the two-sample t test.

6. What is the justification for the non-parametric alternatives to the t tests?

7. What is the reason for the Welch modification to the two-sample t test? What are its relative advantages and disadvantages?

8. What types of problems are handled by the chi-square tests that were introduced here?

9. Explain the importance of power computations. What are the advantages and disadvantages of the Monte Carlo approach to power computation?

10. What does the Kolmogorov-Smirnov (KS) test do that the t tests and its nonparametric competitors fail to do?

Chapter 7
Basic Regression and Analysis of Variance

7.1 Introduction

Early in this book (on page 3) it was claimed that *science usually attempts to describe how such* [measured] *quantities are related to each other.* But up to this point very little has been done here in this area beyond

- some illustrations of simple linear regression in Example 2.9 on page 38, of multiple linear regression in Example 2.10 on page 39 and one-way analysis of variance in Example 2.12 on page 40

- some references to curve fitting in the discussion of scatter plots, starting on page 74

- some discussion of the *sample correlation coefficient,* Definition 3.19 on page 76, and Exercise 5.13.2. on page 178.

The basic reason for this state of affairs is that a background in elementary probability and statistics is essential to any half way decent capability in determining relationships between measured quantities. And this is the background that has been built in the preceding chapters. Even with this background, there are only a few problems that can be handled adequately without expert help by most people who are not highly experienced in this area. One more obstacle is that no statistical package written by human beings could be expected to cover adequately all of the topics in this area that I feel should be included in a first course such as this one.

In light of these difficulties, the material that can be covered in this chapter is very limited, and not adequate to solve some problems which arise early in scientific work. I will, however, indicate what these problems are, and the nature of the solutions that are available to the expert.

7.2 Simple Linear Regression

The first kind of relationship anyone is likely to meet is between two measured numerical quantities, such as between

- calories and weight
- electrical current and voltage
- amount of sunlight exposure and the probability of skin cancer
- automobile speed and mileage.

Over a limited enough range it seems reasonable to approximate such relationships be straight line segments, because our experience with curves indicates that over a short enough stretch, most curves we meet look pretty straight.

If we measure data pairs, such as (automobile speed, mileage), even under very controlled conditions, such as measuring for as close to one hour as is feasible, at a speed as close to 50 miles per hour as it is under our power to do, such repeated independent identical experiments rarely give the same mileage for each run. So such data pairs do not usually lie on a straight line. Since we believe this deviation from straightness is due to factors which we are unable to control, and which still give rise to statistical regularity, it is reasonable, for a short enough range of speeds, to use as a model the formula

$$m = i + sS + error, \qquad\qquad 7.1$$

where i and s are constants (representing intercept and slope, respectively) and S represents the vehicle's speed. Since the range of speeds is assumed reasonably limited, we assume the *error* is normally distributed with mean 0 and unknown standard deviation σ. The normality assumption is questionable, but often provides reasonable answers. Figure 2-10 on page 44 illustrates what line might result from fitting a line to the data points represented by the circles.

The most common method of fitting curves represented by a formula with coefficients to be determined, is *least squares*. Least squares fitting chooses the curve of the specified form yielding the smallest *mean squared vertical deviation* of the curve from the data. By this is meant the following.

Definition 7.1: Least squares curve fit to data
Simple linear regression

Suppose we have data points,[1] (x_i, y_i), and a class of curves or surfaces (sets of points with coordinates (x,y)) given by a formula of the form

$$y = f(x; c_1, ..., c_k).$$

One curve in this class is specified by each choice of the constants $c_1, ..., c_k$.

The *least squares curve for the given data* is the curve in this class whose coefficients $c_1, ..., c_k$ yield the smallest value of

$$\sum_{\text{all } i} [y_i - f(x_i; c_1, ..., c_k)]^2 \qquad\qquad 7.2$$

When the conditions that $k = 2$, x is a number and $f(x; c_1, c_2) = c_1 + c_2 x$ are all satisfied, the least squares curve is called *simple linear regression*.

❑

The formula used in Equation 7.2 was chosen for two reasons

- it is mathematically easy to deal with
- it provides reasonable answers most of the time.

---

1. The points x_i may even be vectors, but the y_i are assumed to be numbers.

Example 7.1: Simple regression on mileage data

We now generate some mileage data, based on the formula $m = 40 - 0.3s$, where s is the speed in miles per hour. Assume we make measurements spaced 3 miles per hour apart, starting at 45 mph, and going to 75 mph, and that the error is normal with mean 0 and standard deviation 4 miles per gallon. Here are the commands and some of the generated data.

```
MTB > set c1
DATA> 45:75/3
DATA> end
MTB > let c2 = 40 - .3*c1
MTB > random 11 c4;
SUBC> normal 0 4.
MTB > let c3 = c2 + c4
MTB > |
```

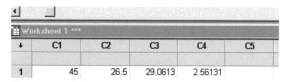

↓	C1	C2	C3	C4	C5
1	45	26.5	29.0613	2.56131	

Now we plot the data, using plot c3*c1, to get

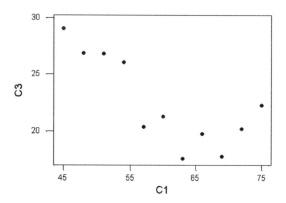

Then we choose Stat -> Regression -> Fitted Line Plot from the menu bar to see the following, which we filled out as shown.

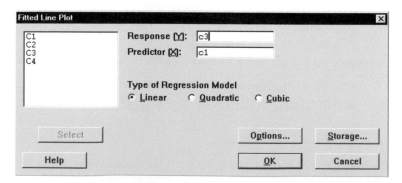

After pressing OK, we obtain

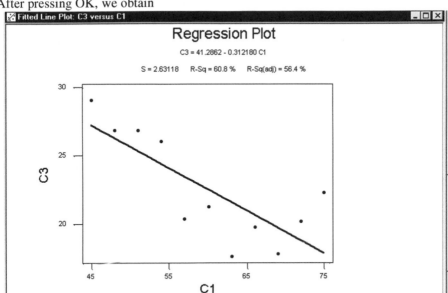

This gives us some information, but we need a little bit more, so instead of choosing Fitted Line Plot, we choose Regression in its place, which yields

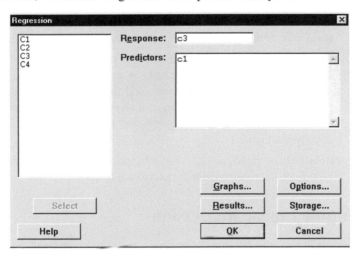

Filling this out as shown, and pressing OK gives these results on the session screen. (Actually there was a lot more, which is not needed.)

Regression Analysis: C3 versus C1

```
The regression equation is
C3 = 41.3 - 0.312 C1

Predictor        Coef      SE Coef         T        P
Constant       41.286        5.080      8.13    0.000
C1            -0.31218      0.08362     -3.73    0.005

S = 2.631       R-Sq = 60.8%     R-Sq(adj) = 56.4%
```

Remember, the actual equation generating this data was $m = 40 - 0.3s$, so these results are pretty good. The label SE Coef is read *standard error of the coefficient*, and is an estimate of the standard deviation of the estimate of the coefficient. From this output we can obtain confidence intervals for the true coefficient (via the t-distribution confidence interval for the mean, see Exercise 5.6.1. on page 158). Be careful, because the listed T value in the output just presented is of no use to you, although SE Coef is what you want to use for the sample standard deviation. Because two parameters were estimated, the lookup table to use for the confidence intervals for the parameters is for the t distribution with n -2 degrees of freedom. If, say, you want to be 95% sure that both confidence intervals will include the true value they are supposed to, get 97.5% confidence intervals for each coefficient (justified by Bonferroni's inequalities, Theorem 4.8 on page 103).

Finally, in the output, the first term on the last line gives an estimate of the standard deviation of the raw data, here 2.631; not perfect, but the right order of magnitude. The values Rsq are of no use if the predictor variables are not random variables. The estimated raw standard deviation, together with the line plot, give a much better idea of how good the model is.

■

Practice Exercises 7.2

1. Repeat the computations of Example 7.1 just presented, but choosing the *quadratic option.*

2. In place of the formula of Example 7.1 above, change the standard deviation to 2, and use the formula $m = -15 + 2s - 0222s^2$ and repeat this example with both the linear, and then the quadratic option. (Note that this is quadratic regression, not **simple** linear regression. It is still linear, because the unknown regression coefficients occur *linearly*, i.e., as simple weights.)

3. Referring to Example 7.1, for what size standard deviation does the quality of the results seem to deteriorate badly?

7.3 Multiple Linear Regression

The multiple linear regression is handled in almost the same manner as the simple linear regression. The basic difference is that there are more predictor variables; that is, x is a vector. So you might want to model the effect of speed and outdoor temperature on mileage. Rather than dream up some formula for this, try some out yourselves, making them mirror what you believe the response to both these variables might be. Then generate some data, just as was done in the simple linear regression, and see how well the equation can be estimated. If you want confidence intervals for the unknown so-called *regression coefficients*, $c_1, ..., c_k$, when all of the regression coefficients can be estimated, the degrees of freedom used is $n - k$ (k being the number of unknown regression coefficients, $c_1, ..., c_k$).

The model is that of independent observations,

$$y_i = f(x_i; c_1,...,c_k) + error_i \qquad\qquad 7.3$$

where

- x_i is a k - dimensional vector which can be under the experimenter's control, the regression coefficients, $c_1,...,c_k$, are assumed unknown, with

$$f(x_i; c_1,...,c_k) = \sum_{j=1}^{k} x_{ij}c_j$$

- the error terms are assumed to be statistically independent, normal with 0 mean and unknown standard deviation, σ. Be careful, because if the assumed constancy of standard deviation is seriously violated, the estimates can be seriously untrustworthy.

If the coefficient, x_{ij}, represents the amount of the j-th medication that is given to the i-th patient, then the regression coefficient would be the average response per unit dose of the j-th medication.

Some of the highly sophisticated computer packages allow tests and confidence intervals about any weighted sums of the regression coefficients. Unfortunately, they tend to be somewhat unwieldy to use.

7.4 The Analysis of Variance

In multiple linear regression the deterministic part of the model (that part of the model not involving variables exhibiting randomness) is usually some algebraic formula[1] which is a weighted sum of the regression coefficients, using known weights. By contrast, the terms of the deterministic part of an analysis of variance model seem to be *indexed variables* and maybe a constant, as for example the one way layout

$$y_{i,j} = a_i + error_{i,j} . \qquad\qquad 7.4$$

Here a_i might represent the average response to the *i-th* treatment. Actually, we can write the equality above so that it looks exactly like a multiple linear regression

---

1. It might be something like $x_1c_1 + exp(x_2)c_2 + \cdots$. This still represents **linear regression**, since the deterministic part of the model is a sum of terms each of which is the product of a factor whose value is known, times an unknown regression coefficient.

model, namely, if say, the index i can take on the values 1,2, or 3, and for each i, there were 3 observations, we may write Equation 7.4 as

$$y_1 = 1a_1 + 0a_2 + 0a_3 + error_1$$
$$y_2 = 1a_1 + 0a_2 + 0a_3 + error_2$$
$$y_3 = 1a_1 + 0a_2 + 0a_3 + error_3$$
$$y_4 = 0a_1 + 1a_2 + 0a_3 + error_4$$
$$y_5 = 0a_1 + 1a_2 + 0a_3 + error_5$$
$$y_6 = 0a_1 + 1a_2 + 0a_3 + error_6$$
$$y_7 = 0a_1 + 0a_2 + 1a_3 + error_7$$
$$y_8 = 0a_1 + 0a_2 + 1a_3 + error_8$$
$$y_9 = 0a_1 + 0a_2 + 1a_3 + error_9$$

Here

$y_{1,1}$ is replaced by y_1, $y_{1,2}$ is replaced by y_2, $y_{1,3}$ is replaced by y_3

$y_{2,1}$ is replaced by y_4, $y_{2,2}$ is replaced by y_5, $y_{2,3}$ is replaced by y_6

etc., and similarly for the error terms.

The main difference is that the known weights occurring in these formulas are either 0 (indicating absence of this term) or 1. Mathematically the analysis of variance model is treated the same way as any multiple linear regression model.

In the analysis of variance model, each indexed variable represents a sequence of distinct regression coefficients, and each of these regression coefficients must be estimated separately. The error terms are assumed statistically independent, and normally distributed with mean 0 and unknown standard deviation σ.

In the following examples we will illustrate how to use the one- and two-way layouts, Tukey's and Dunnett's methods.

Topic 7.2: The one-way layout

The one-way layout is the natural extension of the pooled variance independent two-sample t test. The model is usually written in one of two standard ways:

$$y_{i,j} = a_i + error_{i,j} \qquad\qquad 7.5$$

and

$$y_{i,j} = m + \alpha_i + error_{i,j} \quad \text{with } \alpha_1 = 0 \qquad\qquad 7.6$$

$i = 1, 2,...I, j = 1, 2, J_i$. The only purpose of the second form is that, if say the $y_{i,j}$ represents the measured response of the j-th person in the i-th treatment group, the α_i would represent differences between the i-th group average treatment response, and that of the control group (here chosen as group 1).

There are two ways to store the measured responses,

unstacked, meaning that each group has its own response column and

> stacked, meaning that all of the responses are put in a single column, and the corresponding groups are put in another column (best if it is adjacent to the response column).

For trials with many different treatments, the stacked format is preferable.

We illustrate the unstacked approach, leaving the stacked approach for a later example.

Example 7.3: Comparing three treatments

Here we assume there are 12 people in the first group, 14 in the second group and 11 in the third group. The data, representing recovery times, was generated as normal with means 11.1, 9.2, and 12.4 respectively, and standard deviation 3.1. We put this data in columns c1, c2, c3; we then choose from the menu bar

Stat -> ANOVA -> One-way [Unstacked],   all of the above yielding the following Session screen, worksheet and dialog box.

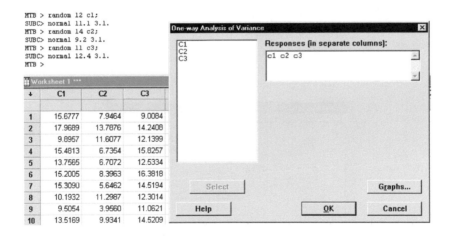

Filling out this dialogue box as shown, and pressing OK yielded the following.

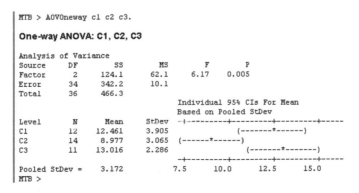

The default null hypothesis that is tested for one-way in Minitab is that the means of the groups are all the same. The default alternative is that they are not all the same. In this case, the low p-value indicates that the null hypothesis would be rejected at any significance level above .005. The graphed confidence intervals provide some idea of how well the three means are known using the given data.

The confidence intervals could have been computed from the sample means and standard errors (estimates of the standard deviations of the sample means), using the t based confidence intervals of the same type used in Example 7.1, which started on page 219, but with $n - 3$ degrees of freedom because 3 parameters are being estimated.

Unfortunately the Minitab one-way ANOVA does not supply the most accurate information about the magnitude of differences of these means. If you had wanted this information, rather than run one-way, you could use Tukey's method for looking at all pairwise differences, or Dunnett's method for comparisons only with the control group. To do either of these, you must use the stacked version of one-way. So before starting, in Minitab type   stack c1 c2 c3 c5, followed by putting a 1 in the first 12 rows of c6, a 2 in the next 14 rows, and a 3 in the next 9 rows (to indicate that the first 12 elements of c5 were from group 1, the next set of 14 elements were from group 2 and the last 11 elements from group 3). Then choose from the menu bar as follows:

Stat -> ANOVA -> One-way   to get the following dialog box.

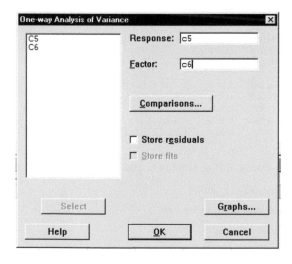

Fill it out as shown, noting that the factor column entries show which group each observations comes from.

In order to get either the Tukey or Dunnett comparisons, click on the Comparisons box, and then click on OK, to get the following dialogue box.

One-way Multiple Comparisons

☑ <u>T</u>ukey's, family error rate: 5

☐ <u>F</u>isher's, individual error rate: 5

☑ <u>D</u>unnett's, family error rate: 5

<u>C</u>ontrol group level:

☐ <u>H</u>su's MCB, family error rate: 5

  ⦿ <u>L</u>argest is best
  ○ <u>S</u>mallest is best

| Help | <u>O</u>K | Cancel |

We fill in the dialogue box as shown (the 5 that was entered indicates that we require that there only be a 5% chance that even one of the intervals supplied is incorrect). Pressing OK, we get the following response from the computer, where the first part shows the Tukey results, and the remainder the Dunnett results. (Actually the entire one-way ANOVA was also printed, but one should use either the Tukey, the Dunnett, or the one-way, but not more than one of these programs. In particular, if you are really interested in all pairwise differences, you should use only Tukey's method. Dunnet's or Tukey's method is not meant to be applied after running an ANOVA, although this is commonly done.

```
Dunnett's comparisons with a control

    Family error rate = 0.0500
Individual error rate = 0.0273

Critical value = 2.31

Control = level (1) of C6

Intervals for treatment mean minus control mean

Level    Lower    Center    Upper  --+---------+---------+---------+-----
2       -6.363    -3.485    -0.606  (--------*---------)
3       -2.500     0.555     3.609                  (---------*---------)
                                    --+---------+---------+---------+-----
                                    -6.0      -3.0      0.0       3.0


Tukey's pairwise comparisons

    Family error rate = 0.0500
Individual error rate = 0.0194

Critical value = 3.47

Intervals for (column level mean) - (row level mean)

                  1              2

       2        0.422
                6.547

       3       -3.804          -7.176
                2.695          -0.903
```

If you have trouble interpreting the Tukey output, the interval for $\mu_i - \mu_j$ is the interval (written vertically so that the left endpoint is the upper of the 2 numbers) is the pair of numbers in row i (which can only be 2 or 3) column j.

∎

A strange property of ANOVAs in general, which is well illustrated by the one-way ANOVA. Suppose you have k means, $\mu_1,...,\mu_k$, and you want to test the hypothesis that all of the differences $\mu_1 - \mu_2, \mu_2 - \mu_3,...,\mu_{k-1} - \mu_k$ are 0. If you had the actual means, you could look at the individual differences, and see directly if any of the differences were 0. Alternatively you could say, look at all sums of the form $a_1(\mu_1 - \mu_2) + a_2(\mu_2 - \mu_3) +\cdots+ a_{k-1}(\mu_{k-1} - \mu_k)$, i.e. for all choices of the a_i. All of the differences of the means are 0 if and only if all of these sums are 0. Because clearly if all of the differences $\mu_{i-1} - \mu_i$ are 0, then all of the above sums must be 0. But if all of the above sums are 0, by choosing $a_1 = 1, a_{i-1} = 1$, and all other a_j to be 0, we find $\mu_{i-1} - \mu_i = 0$.

In essence, the one-way ANOVA test looks at what the data tells us about all of the sums to draw its conclusions If it chose $a_i = 1$ for $i = 1,...,q - 1$ and $i = q + 1,...,k$ and $a_q = 0$, and found **that sum** to be nonzero, then by the reasoning we just went through, at least one difference $\mu_{i-1} - \mu_i \neq 0$. But what it found from that sum was that $\mu_1 - \mu_q + \mu_{q+1} - \mu_k \neq 0$, which certainly does not tell us which difference $\mu_{i-1} - \mu_i$ differed from 0.

Many statisticians and scientists still think that just because the one-way ANOVA rejected the hypothesis that all the means are the same, they can find which means are different. This is similar to the belief that if you found a change of 2 inches in a child's height over a one year period, and therefore knew that on at least one day in that year the child grew, you could pinpoint such a day. No one would seriously believe this.

The so-called post hoc tests (and there are many) purport to do this, but they can't. If your main interest when you did the one-way ANOVA was to find which means are different, you should omit the one-way ANOVA, and simply run the Tukey procedure. If you were mainly interested in differences from control, just run Dunnett's procedure.

Topic 7.3: The additive two-way layout

The additive two-way layout is a generalization of the paired t test (see Topic 6.4 on page 190) to the situation where there are more than the usual *pre* and *post* observations associated with a given individual. This model is often used when a fixed number of measurements related to some treatment is made on each treated patient. The model is of the form

$$y_{s,t} = m + S_s + T_t + error_{s,t} \qquad 7.7$$

where S_s is the average effect of subject s on the response, $y_{s,t}$, and T_t is the average effect of treatment t on the response. In most packages, $s = 1$ represents the control subject, with $S_1 = 0$, and $t = 1$ represents the control treatment with

$T_1 = 0$. Under these assumptions, m represents the average response of the control subject under the control treatment.

The purpose of putting a subject effect term in this model is to try to reduce the variability due to fluctuations arising from varying characteristics of different subjects. For instance, many measurements might be made on each patient given some specified amounts of blood pressure reduction medication. If each amount were given only once to any patient, the variability in the measurements would likely be very large, because the patient to patient variability would tend to mask the effects of the medication.

The two default tests are usually

> that all treatments have the same effect — i.e., the for all t, $T_t = 0$, and

> that all subject effects are the same.

The latter test is not of much interest here. However when the second effect is, say, the effect of an additional treatment, it makes good sense to be able to test both treatments.

Example 7.4: Blood pressure medication

Assume we have four subjects, each given various doses of some blood pressure medication. For some odd reason, we want to see if these doses all have the same effect[1]. The Minitab specialized two-way ANOVA program requires a balanced design, meaning the each patient takes each dose the same number of times.

We assume that there are three possible doses, and suppose that

$$T_2 = 10 \quad \text{and} \quad T_3 = 13.$$

The 12 responses will be put in column c1. The patient number will be put in column c2 , the first patient being in rows 1-3, the second in rows 4-6, etc. The corresponding dosage index will be put in column c3. We fill column c1 by putting the average blood pressure prior to treatment in the appropriate positions in column c5 (the subject, or patient effect), the average dose effect in the appropriate positions in column c6,and finally the random error (normal with mean 0 and standard deviation 2.5) in column c7. The patient effects were chosen arbitrarily, but could also have been randomly chosen, separately for each patient. The standard deviation of the error was taken as 2.5. The measurement (response) represents the blood

---

1. Actually, the reason may not be that odd, since it seems to have applied to finding
   the minimum effective aspirin dosage to prevent heart attacks.

pressure of the patient after each treatment dose. Column c1 is the sum of columns c5, c6, and c7. All of this shows up in the screen snapshot which follows.

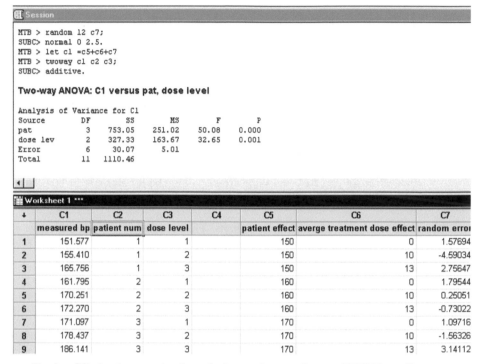

```
MTB > random 12 c7;
SUBC> normal 0 2.5.
MTB > let c1 =c5+c6+c7
MTB > twoway c1 c2 c3;
SUBC> additive.
```

Two-way ANOVA: C1 versus pat, dose level

```
Analysis of Variance for C1
Source       DF       SS       MS        F       P
pat           3    753.05   251.02    50.08   0.000
dose lev      2    327.33   163.67    32.65   0.001
Error         6     30.07     5.01
Total        11   1110.46
```

↓	C1	C2	C3	C4	C5	C6	C7
	measured bp	patient num	dose level		patient effect	averge treatment dose effect	random error
1	151.577	1	1		150	0	1.57694
2	155.410	1	2		150	10	-4.59034
3	165.756	1	3		150	13	2.75647
4	161.795	2	1		160	0	1.79544
5	170.251	2	2		160	10	0.25051
6	172.270	2	3		160	13	-0.73022
7	171.097	3	1		170	0	1.09716
8	178.437	3	2		170	10	-1.56326
9	186.141	3	3		170	13	3.14112

Having filled columns c1, c2, and c3, we choose   Stat -> ANOVA -> Two-way

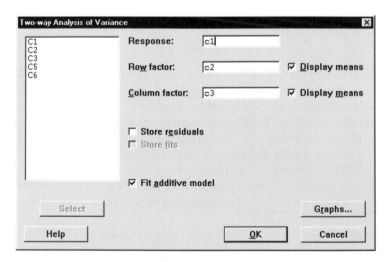

fill out as shown, and press OK, to obtain the following output on the session screen.

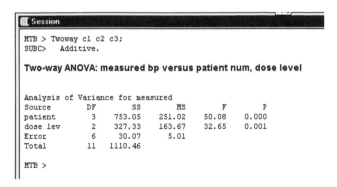

This output indicates that the hypothesis that the treatments all had the same average effect is rejected at the .05 significance level, because the p-value for c3 (dose level) is .001.

■

Topic 7.4: The general two way-layout

The formula for the general two-way layout model is

$$y_{s,t} = m + S_s + T_t + (ST)_{s,t} + error_{s,t} \,. \qquad 7.8$$

The extra term, $(ST)_{s,t}$, is a so-called interaction term, and indicates an extra effect at levels s and t over and above what would be expected by looking at the two factors separately. It is useful for assessing the result of combining drugs, such as alcohol and barbiturates, the effect on reaction time being more than would have been obtained by just adding their separate effects. In this model, in addition to the constraints

$$S_1 = 0, T_1 = 0 \,,$$

it is necessary to require additional constraints, usually

$$(ST)_{1,t} = 0 \quad \text{for all } t, \text{ and} \quad (ST)_{s,1} = 0 \quad \text{for all } s.$$

This is just as easy to run, the only difference being that you do not choose the additive model. In Minitab, if each combination of (patient, treatment) occurs only once, the additive model is chosen anyway, since under these circumstances there isn't enough data to estimate all of the unknown parameters. That is, Minitab just assumes that the interactions are all 0 when there is not enough data to estimate them properly. Note that this is the case here because there are only 12 observations, and 12 regression coefficients to be estimated: m (1), the S_s (3 because the control is set at 0 and there are 4 patients), the T_t (2 because the control is set at 0 and there are 3 doses), and the $(ST)_{1,t}, (ST)_{s,1}$ (12 - [2+3] = 7, because there are 4x3 (ST) terms, of which 5 are 0). But with the same number of parameters to be estimated as there are observations, the model would fit the data perfectly, and there would be no reasonable way to estimate the standard deviation, σ, and we assume there is a nontrivial random component in this model. So Minitab chooses a simpler model with fewer parameters.

You will notice, if you do the first two exercises in the set to follow, that if there are sufficiently many observations, you can use the resulting ANOVA output to test the hypothesis that the interactions are 0, as well as testing that the treatment effects are 0. But be careful about the interpretation, because if you conclude that the treatment effects are 0, this only means that the terms T_t are 0. If the interactions are not 0, then there is certainly an effect from some of the treatments on the observed data. Restated: you can only conclude that the treatments have no effect if both the treatment effects and the interaction effects are 0 in the general two-way layout model.

Practice Exercises 7.5

1. Try running the general two-way ANOVA on the example just presented. See what you determine about the interactions (which were not supposed to be present in the example given).

2. Instead of taking 12 observations, as in the previous exercise, double that number using the same method to generate the second group of 12 as was done for the group already done. Now note the difference between the outputs of the additive model and the general two-way model. In particular, use the output to test the hypothesis that there are no *real* interactions.

3. Try running the nonparametric version (Kruskal-Wallis test, under Nonparametrics) of the one-way ANOVA on the data generated as in Example 7.3 on page 224, to see how the results compare.

4. Try running a one-way ANOVA on data like that in Example 7.4 on page 228, to see how not taking account of subject effects can deteriorate the results that statistical analysis can give.

This chapter was only a very general introduction to a subject to which many semesters are devoted in most statistics graduate programs. To attain a reasonable amount of competence in this area requires (in my view) at least a full year course with accompanying *hands-on* experience.

Epilogue

This text, being an introductory one, made no attempt at completeness. Though it is nowhere near as vast a subject as, say chemistry, there is little hope for any one individual to keep up with all the parts of the fields of probability and statistics. Some of the more important areas that are extensions of what was covered here are the following.

- The theories of testing hypotheses and estimation
- Linear modeling, Generalized Linear Models, Weighted Least Squares, Nonlinear Regression, Logistic Regression (all of which permit much more refined regression type models)
- Random Processes, which study statistically dependent series of observations
- Time Series Analysis, the statistical analysis of Random Processes
- Sequential Analysis, which eliminates having to choose the sample size in advance of seeing the data

These topics, and undoubtedly others, including the mathematical background needed to develop these areas, are worth learning by those who want a fuller understanding of the tools they depend on.

Some final words about the nature of statistics.

While moderately accurate estimation is usually possible with samples of reasonable size, it is usually not realistic to expect to achieve great accuracy by just taking a sufficient number of observations. Because (as can be seen by looking carefully at the Central Limit Theorem), the error of estimation as a function of *raw sample size* is usually on the order of magnitude of $1/\sqrt{\text{sample size}}$. So to achieve an error reduction of 10 times what you have with your current sample size, you would need 100 times the data. Besides the exorbitant cost, there is the real danger of getting careless with so much data. A better approach than bludgeoning the problem with more of the same data is to look for other types of data that will contribute a greater return for the effort expended. This seems especially to be the case if it is the case that much of the random error arises from factors that are being ignored, rather than solely from factors which are too difficult to measure.

Finally, it is necessary to remember that almost all statistical models are approximate descriptions, which have limitations. These models are not worthy of blind unquestioning faith, but many of them can lead us to better predictions and more adequate tools for interpreting the data that we are likely to produce.

Bibliography

[1]   J.R. Blum and Judah Rosenblatt (1972). *Probability and Statistics*, W.B. Saunders Company, Philadelphia, PA.

[2]   Michael R. Chernick (1999). *Bootstrap Methods, A Practitioner's Guide* John Wiley and Sons, New York, N.Y.

[3]   Bradley Efron and Robert J. Tibshirani (1998). *An Introduction to the Bootstrap,* CRC Press LLC, Boca Raton, FL.

[4]   W. Feller (1968). *An Introduction to Probability Theory and its Applications*, Vol. 1, John Wiley and Sons, New York, N.Y.

[5]   George S. Fishman (1996). *Monte Carlo Concepts, Algorithms and Applications*, Springer Verlag, New York, N.Y.

[6]   J.G. Kalbfleisch (1985). *Probability and Statistical Inference*, Vol. 1,2, Springer Verlag, New York, N.Y.

[7]   C. Radhakrishna Rao and Helga Toutenburg (1999). *Linear Models*, Springer Verlag, New York, N.Y.

[8]   Judah Rosenblatt (1961), "Statistical Aspects of Collision Warning System Design Under Assumption of Constant Velocity Courses," *Naval Research Logistics Quarterly*, 8:4, 317-341.

[9]   Judah Rosenblatt and Stoughton Bell (1999). *Mathematical Analysis for Modeling*, CRC Press LLC. Boca Raton. FL.

[10] J.V. Uspensky (1937). *Introduction to Mathematical Probability,* New York, McGraw Hill Book Company, New York, N.Y

Selected Answers and Solutions

Px stands for Practice Exercise, Prob stands for Problem

Chapter 1

Prob 1.1.1 Yes. Fill up the tank with gasoline. Note the odometer reading. Drive in your normal pattern (assuming that wherever you are, you can get to a gas station with the amount of fuel remaining). When your tank registers 1/4 full, head to a gas station and fill up.

Compute $\dfrac{\text{miles driven}}{\text{amount of gasoline to fill up}}$. This is your mileage estimate.

Prob 1.1.4 No. Because you have little assurance that you would get proper information from sociopaths (people who lie, but show little physical evidence of it).

Prob 1.1.6 The real issue is whether it is more convenient to phrase problems in terms of the sun as coordinate system origin, or the earth as coordinate system origin. Either can be used to determine the time of solar eclipses, but the first is far simpler to use.

Prob 1.1.7 Even in theory this problem doesn't seem to be amenable to solution, because the structure of society, the educational system, and peoples' attitudes seem to be constantly changing.

Prob 1.2.1 Probably yes. The variables might be:

- Latest acceptable arrival time
- Day of the week
- Month of the year
- The weather
- Amount of time adequate for dressing and breakfast

Prob 1.2.4 The answer appears to be yes. CDC must determine suitable descriptions of what flu strains are prevalent for the given year, where and when they will originate, and how they are likely to spread. Then they must develop vaccines for those strains likely to affect sufficiently many people. For all of these they will need models — problems beyond the scope of this question.

Prob 1.2.5 This is a question, that if answered poorly, could create real problems for GM. Some important variables are

cost of development of reliable electric brakes,

their cost of manufacture (under mass production)

the price people would be willing to pay,

the cost of user maintenance.

Prob 1.5.1 Use a deterministic model if the deviations from the (deterministic) answer can be tolerated in applying it to the situation of interest. Otherwise try for a statistical model whose predictions are adequate — there may not be any suitable model available.

Prob 1.5.2 A deterministic model, that takes proper account of skin color should be adequate for this purpose. A statistical model might be somewhat more convenient for judging adequacy in specific situations.

Prob 1.5.5 A deterministic model, used conservatively, would probably be adequate, although a statistical one would appear more convenient.

Px 1.4.1 Each triple (d_1, d_2, d_3), $d_i = 1, 2, 3, 4, 5, 6$ is assigned probability $1/216 = 1/6^3$. The number of triples summing to 11 can be written as

sd10 + sd9 + sd8 + sd7 + sd6 + sd5 = 3 + 4 + 5 + 6 + 5 + 4 = 27,

where sdj is the number of pairs yielding a sum of j.

So P{sum of 11 on 3 tosses of a fair die} = 27/216 = 1/8 = .125

Px 1.4.3 The code changes to the following:

First we will make 100,000 tosses (rather than 1000) for better accuracy.

Next we will fill three columns (rather than two) to simulate three tosses. That's about it.

Here is the code.

```
random 100000 c1-c3;

integer 1 6.

let c4 = c1 + c2 + c3

let c4 = c4 - 11

let c5 = round(c4/(c4 + .001))
```

let c6 = 1 - c5

mean c6

The result we got was .12409 — pretty good.

Prob 1.7.2 To better compare different procedures by applying them to the same data — so that the differences you see reflect the change in procedure, rather than the changes in the data. This is similar to the use of matched pairs (matched on gender, age, etc.) to compare treatments (see Topic 6.4 on page 190).

Prob 1.7.8 For very long run predictions it would seem to me that changes over which human beings have no control or predictive ability would make even the choice of the form of a model into an insuperable problem. However, we could probably, in theory, reasonably determine a class of models, one of which is suitable to describe what happens in response to whatever actions human beings may take over a short period of time (e.g., 50 years). However, to run a controlled experiment which would give us faith in the chosen model for prediction, would probably not be feasible — considering that agreement on issues of far less importance is so hard to come by.

Prob 1.7.10 The variables that need to be characterized from ideal data might include

refined batting statistics of each player against each opposing pitcher (i.e., P{single}, P{double},... etc.
runners' stolen base records

The manager could then experiment to determine best batting order using Monte Carlo techniques. If a pitcher fatigue factor was included in some way, the question of when to put in a relief pitcher could be addressed.

Chapter 2

Px 2.1.2

a. If you are able to print out all four graphs with the same x and y scales (and this takes a bit of work) you will see that the further away p is from .5, the more rapidly the graph seems to settle down to the value p. This is because the further p is from .5, the fewer jumps per trial it is likely to have.

b. Since the sum of the 2 graphed functions is 1, the answer is essentially the same.

Px 2.4.1 The computed value is .0796. (I take the Fifth Amendment on my own guess.)

Px 2.6.4 The difficulty with this exercise is that when we desire to show all observances in the graph, the outliers (observations whose distances from most of the other observations have large magnitudes) tend to make the units on the x axis so small that we cannot see what is going on. One way to handle the problem is to trim off (eliminate) some percentage of the largest and smallest values. We will trim the top and bottom 4%. We first generate 2000 Cauchy variables (similarly to what was done in Example 2.3 on page 28) and delete the smallest 80 values and the largest 80 values by respectively *selecting these values* (hold down the shift key while using the *down arrow*) and then pressing the *Delete* key. We would also multiply the Cauchy density that fits over the remaining range of values by 2000/(2000-160)=2000/1840, via the command line   let c3 = c3*2000/1840, so that its total area over this range is approximately equal to 1. The result is shown below.

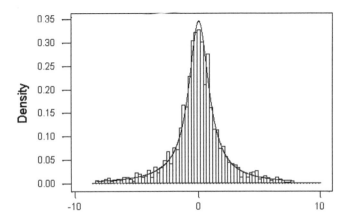

Prob 2.16.1 Using Calc -> Probability Distributions -> Binomial, we find

a.

p	$B(501,1001,p)$	$1 - B(501,1001,p)$
.46	.9953	.0047
.47	.9752	.0248
.48	.9082	.0918
.5	.5252	.4748
.52	.1145	.8853
.53	.0331	.9669
.54	.0067	.9933

b. Pretty well.

c. Since $B(3001,6001,.49) = .9424$ and $B(4001,8001,.49) = .965$, we see that the required sample size is somewhere between 6001 and 8001. With some tedious additional work the exact sample size can be found.

Prob 2.16.4 If the model is $y = a + bx + error$ and the question is *Does a change in x produce a "real" change in y?* the simple linear regression test *Does a = 0?* addresses this question. On the other hand, if the data consisted of five observations at each $x = 0, .01, .02,...,9.99, 10$ (5005 values) this data could be modeled by $y_{ij} = \mu_i + err_{ij}$ and testing the hypothesis that all μ_i are the same value would also address this question. The ANOVA formulation would be a very inefficient way to answer this question if the data was known to arise from a straight line, because for the ANOVA model to be uniquely defined requires knowledge of 1001 parameters μ_i $i = 0,1,...,1000$, all of which must be estimated, while the regression model has only two parameters (a and b) to be estimated. The *cost* of estimating many parameters (in terms of how well the test performs) is quite large. But the ANOVA class of models is more general than simple linear regression, in the sense that it can describe a greater variety of curves.

Prob 2.17.2 Divide the interval from 0 to 1 into subintervals, the i-th one having length p_i. If the value of the uniform random variable that is observed lies in the i-th interval, the value of the random variable being generated is taken to be i. That is, $P\{X = i\} = p_i$. This idea can be extended to use uniformly distributed random variables to generate random variables with any distribution.

Chapter 3

Px 3.1.2 The attribute might be the existence of the HIV virus in the bloodstream. Or it could be the existence of antibodies to this virus in the bloodstream, or even the existence of some toxins (poisons) associated with the HIV virus.

Px 3.1.3 Here the attribute of interest would be p_{800}, the probability that light bulbs of this type will burn (produce acceptable light) for at least 800 hours.

Px 3.3.2 In my sample there were 11 0s, 11 1s and 3 2s.

Px 3.3.3 The sample mean was 6.9724 (close to 7).
The minimum value was 0, the maximum was 16.

Prob 3.8.2 Here you fill column c1 with the values 1 to 900 (via Calc -> Patterned Data, following the directions appropriately to fill column c1).

After clicking OK, type sample 900 c1 c2.

Use the first 300 rows of c2 for the first group, the next 300 rows of c2 for the second group and the last 300 rows of c2 for the last group.

Next select the first 300 rows of c2 (click on the first row of c2, hold down the shift key and use the down arrow) and put them in the first 300 rows of c4; select the next 300 rows of c2 and put them in the first 300 rows of c5, and put the last 300 rows of c2 into the first 300 rows of c6 (You can use Edit -> copy   after selecting, and then use Edit -> paste after clicking in the appropriate cell to do this moving of data.). After this, do a describe of c4-c6 to see if the groups seem similar. Here were the results I got that illustrate their similarity.

Descriptive Statistics: C4, C5, C6

Variable	N	Mean	Median	TrMean	StDev	SE Mean
C4	300	445.4	447.0	444.4	256.5	14.8
C5	300	446.5	422.5	446.1	262.7	15.2
C6	300	459.6	470.0	461.1	261.3	15.1

Variable	Minimum	Maximum	Q1	Q3
C4	3.0	900.0	221.8	654.8
C5	1.0	896.0	225.3	686.8
C6	4.0	899.0	227.8	697.5

Prob 3.9.2 In our simulation we typed

```
random 100 c1-c1112;
bernoulli .5.
rmean c1-c1112  c1114
let c1 = c1114
sort c1 c1
```

We detected exactly three rows of column c1 outside the range (.47, .53). With .5 replaced by $p = .2$ there were no rows of column c1 outside the range (.17, .23), making the assertion pretty believable.

Px 3.7.2 The sample of size 2 going across rows, consisted of the numbers 660 and 496. (Repeats would have been tossed out.) Not reusing the tabled values already examined, the sample of size 3 consisted of 071, 895, and 444. For the sample of size 975, we first chose a sample of size 5 from the population of size 980 and excluded these values from the original 980 — leaving the sample of size 975 that we wanted. The five excluded values were 018, 512, 293, 789, and 110. Note that we just wanted three samples (nothing said about whether there were any common members).

Prob 3.10.2 First we put 0,1,2 in rows 1,2, and 3 respectively of column c1. Then we put .49, .01, and .49 similarly in rows 1,2, and 3 of column c2. Next we typed

> random 20 c3-c12;
>
> discrete c1 c2.
>
> random 1000 c13-c22;
>
> discrete c1 c2.
>
> random 10000 c23-c32;
>
> discrete c1 c2.

The sample medians of the rows were put in columns c34, c35, and c36 respectively via the commands

> rmedian c3-c12 c34
>
> rmedian c13-c22 c35
>
> rmedian c23-c32 c36

The easiest way to examine the behavior of these three sequences of sample medians is via histograms, of which we got the following:

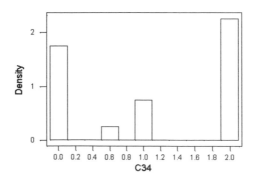

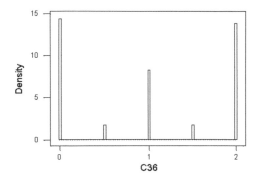

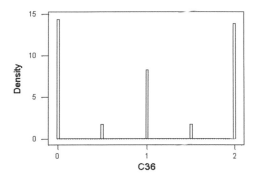

Because there were very few medians in the first group the first histogram could not be predicted very well. (So your first histogram may differ substantially from the first one presented here.) The histogram for column c35 is probably a better than expected illustration of statistical regularity, while the one from column c36 is probably somewhat poorer than usual.

To see how such a state of affairs can happen, if you toss a fair coin twice, there is a 50-50 chance of getting 50% Heads — which is in perfect agreement with the underlying mechanism. If a fair coin is tossed 10,000 times, there is almost no chance of perfect agreement with the underlying mechanism because

$$P\{5000 \text{ Heads in } 10,000 \text{ tosses}\} = .0080.$$

So we see that in two tosses of a fair coin, the proportion of Heads has a 50% chance of differing from P{Head} by .5, while in 10,000 tosses of a fair coin, the probability that the proportion of Heads differs from .5 by more than .01 is less than .0466. These calculations were carried out by choosing as follows from the top menu bar.

Calc -> Probability Distributions -> Binomial .....

Prob 3.10.4 Following the directions, we see that in 69 of the 100 cases, the sample mean based on 40 observations was closer to the population mean based on 20 observations (and further from the population mean in the remaining 31 observations), in our sample. So

$$P\{\overline{X}_{40} \text{ is closer to the population mean than } \overline{X}_{20}\} \cong 0.69.$$

Prob 3.12.2 From a visual examination of the data, the range of the data (the difference between the largest and smallest values) seems to increase as the mean increases. But, again from visual examination, the ratio

$$\frac{\text{largest - smallest}}{\text{sample mean}},$$

which is a measure of the relative error, seems to decrease as the mean increases, as indicated.

$$\frac{18-5}{10} = 1.3 \qquad \frac{41-6}{30} = \frac{5}{6} \cong 0.833 \qquad \frac{125-76}{100} = 0.49.$$

Prob 3.12.4 The tsplot shown below provides no apparently useful information for solving this problem.

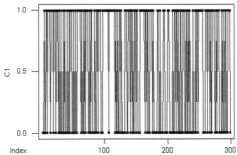

From the meanseq plot shown below, drawing horizontal lines at p = .5, .52, and .54, at 300 observations the final sample mean value was between .50 and .54, although nearby (at 250 observations) it seemed to go above .54.

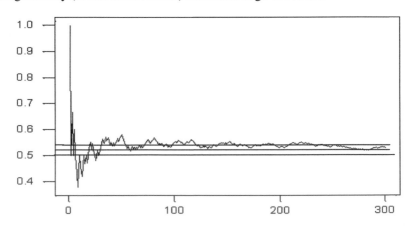

So from this one graph, looking at the variability, it doesn't seem to be definitive that 300 is an adequate sample size. So we now follow the 3rd suggestion. We found 237 sample means below .5 (by sorting the column of sample means to make counting easy). Thus it looks as if we would make the correct decision when $p = .52$ only about 78% of the time. Hence it looks as if we would need more than 300 observations if we wanted to be correct in our predictions 95% of the time, when $p = .52$.

Prob 3.12.6 All that needs to be done is to change the single line rmean c1-ck2 ck3 to   rmed c1-ck2 ck3

Px 3.9.1

   a. all of them ($2\sigma = 0.58 > 0.5$)

   b. The standard deviation of $X_1 + X_2$ is approximately .409.

   c. We want to know the proportion of $X_1 + X_2$ values between -.818 and +.818. Using sort we find the observed proportion of our sample in this region to be $(1000 - 22 - 16)/1000 = 0.972$ (more than 95%)

   d. The standard deviation of $X_1 + \cdots + X_{10}$ is about .903.

   e. From the answer to part d. it follows that we want to know the proportion of values of $X_1 + \cdots + X_{10}$ that lie between -1.806 and +1.806. The number of observations in this region is found (with the help of the sort command) is 1000 -20 -26 = 954. So 95.4% of the $X_1 + \cdots + X_{10}$ values are within two standard deviations of the mean. We conclude that the two-sigma rule of thumb (Topic 3.14 on page 68) is beginning to look accurate for a sum of at least 10 uniform random variables.

Px 3.9.2 First we will estimate the Poisson standard deviation when the mean is 55 by generating 1000 such variables and then using the describe command. The estimated standard deviation turns out to be 7.416. It is known from theory that the true standard deviation for a Poisson variable with a mean of 55 is $\sqrt{55} \cong 7.4162$, so the simulation worked well. Now we will empirically check the two-sigma rule of thumb (Topic 3.14 on page 68) on this data set. We find that the number of observations in the range from 55 - 2σ to 55 + 2σ = 55 - 14.8 to 55 + 14.8 = (40.2, 69.8) is 1000 - 20 - 35 = 945. So the two-sigma rule seems to hold here.

The observed value of 94 is (94 - 55)/7.4 = 39/7.4 = 5.27 standard deviations from the mean. Values as large or larger than this are extremely unlikely. In fact, using Calc -> Probability Distributions -> Poisson from the top menu bar, we find that $P\{Y \geq 94\} = 1$ to at least five significant figures. I would worry.

For the Poisson distribution, the two-sigma rule of thumb seems to hold pretty well for means at least 16 (the observed proportion of values within 2σ of the mean in a simulation of 10,000 values, for a mean of 16 is .9528).

Prob 3.15.2 Throwing away outliers (values far away from most of the other data) would make relative differences in position easier to spot.

Px 3.12.3

   a. Yes. In fact Y is determined perfectly by X.

   b. The sample correlation coefficients were -.021, -.023, -.207, .04, .441.

I claim that this data suggests that the true correlation coefficient is 0, since it certainly does not seem to favor a positive or a negative value.

Since a plot of $y = x^2$ vs x looks as shown and negative x values tend to occur

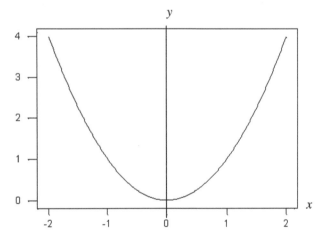

symmetrically with positive ones about 0 (since the distribution was standard normal) you certainly would not expect the algebraic value of x to increase [or decrease] when the algebraic value of y increases [or decreases]. Since the average of any of the two predictions of X from any given Y is 0. So it is reasonable to believe that here the true correlation coefficient is 0 (the mean f the X population) and this is unchanged even if you know the value of Y.

Px 3.13.2 We only present results for the sample size of 60. The normal 0, 1 empirical distribution function and the normal 0, .2 distribution function are very different despite having the same mean, as can be seen from the graph produced by the empdis macro.

Px 3.13.4 We present results only for the sample size of 60. The Poisson hypothesized distribution function and the normal 1 1 empirical distribution function are relatively close to each other, compared with the previously compared pair.

Prob 4.5.2.a $Q_d = \{(x, y) \text{ in } S : xy = d\}$, $Q_0 = L_0 \cup M_0$.

For $d \neq 0$ $Q_d = \underset{\substack{x \text{ real} \\ x \neq 0}}{\cup} (L_x \cap M_{d/x})$.

Q_0 consists of the x and y axes.

If $d \neq 0$ then Q_d looks as the indicated curves below.

$$d > 0 \qquad\qquad\qquad d < 0$$

Prob 4.7.2 The smallest value is achieved if the set with lower probability is a subset of the other one. Then $P\{A \cup B\} \geq max(P(A), P(B))$

Px 4.15.2.a
$$\frac{\binom{18}{3}\binom{82}{17}}{\binom{100}{20}}$$

Px 4.15.3
$$\frac{\binom{13}{3}\binom{13}{2}\binom{13}{5}\binom{13}{3}}{\binom{52}{13}} = 0.01293$$

Px 4.18.1
$$\frac{6/36}{6/36 + 2/36} = \frac{6}{8} = \frac{3}{4} = 0.75$$

Px 4.18.2 The sample space could consist of as few as four elements —

$$(S, L), (\sim S, L), (S, \sim L), (\sim S, \sim L),$$

where S represents that the person being looked at is a smoker,
L represents that the person being looked at gets lung cancer
$\sim S$ represents that the person being looked at does not smoke
$\sim L$ represents that the person being looked at does not get lung cancer.

If probabilities totaling 1 are assigned to these four points, the conditional probability of getting lung cancer given that the person smokes can be found.

Px 4.18.4

a. The theoretical median is the value of t for which $1 - e^{-t/500} = 1/2$, i.e.,

$e^{-t/2500} = 1/2$ or $t = -2500 \ln(1/2) = 1732.87$, where $\ln$ is the natural logarithm function ($\log$ to the base $e = 2.718281828...$)

b. For $T > 0, \tau > 0$ we want

$$P\{T > T + \tau \mid T \geq \tau\} = P\{T > t + \tau, T \geq \tau\}/P\{T \geq \tau\} = P\{T > t + \tau\}/P\{T \geq \tau\}$$

$$= \frac{e^{-(t+\tau)/2500}}{e^{-\tau/2500}} = e^{-t/2500} = P\{T > t\}$$

Px 4.18.6

a. $\dfrac{1}{2^4} = \dfrac{1}{16}$

b. Let S_5 be the total number of successes in the five trials. We want

$$P\{S_5 = 5 \mid S_5 \geq 1\} = \frac{P\{S_5 = 5, S_5 \geq 1\}}{P\{S_5 \geq 1\}}$$

$$= \frac{P\{S_5 = 5\}}{P\{S_5 \geq 1\}} = \frac{P\{S_5 = 5\}}{1 - P\{S_5 = 0\}}$$

$$= \frac{1/2^5}{1 - 1/2^5} = \frac{1}{2^5 - 1} = \frac{1}{31}$$

Px 4.40.2.b

$$P\{\text{dice show the same face} \mid \text{sum is odd}\}$$

$$= \frac{P\{\text{dice show the same face, sum is odd}\}}{P\{\text{sum is odd}\}}$$

We first compute the denominator term. We note that it is easy to show that the tosses are statistically independent, justifying some of the steps below.

$P\{\text{sum is odd}\}$

$\quad = P\{\text{sum is odd, last die is odd}\} + P\{\text{sum is odd, last die is even}\}$

$\quad = P\{\text{sum on 1st 2 dice is even, last die is odd}\}$

$\qquad + P\{\text{sum on 1st 2 dice is odd, last die is even}\}$

$\quad = P\{\text{sum on 1st 2 dice is even}\}P\{\text{last die is odd}\}$

$\qquad + P\{\text{sum on 1st 2 dice is odd}\}P\{\text{last die is even}\}$

The second factors have probabilities 2/3 and 1/3 respectively. The first factor of the first term has the value 4/9 (corresponding to both tosses being odd = (2/3)*(2/3)) + 1/9 (corresponding to both tosses being even = (1/3)*(1/3)). The first factor of the second term must therefore be 5/9. Hence

$$P\{\text{sum is odd}\} = \frac{4}{9}\frac{2}{3} + \frac{5}{9}\frac{1}{3} = \frac{13}{27}.$$

The probability $P\{\text{dice show the same face, sum is odd}\}$ is just $P\{(1,1,1), (3,3,3)\}$ which is 2/27. Hence

$$P\{\text{dice show the same face} \mid \text{sum is odd}\} = 2/13.$$

Px 4.20.3 Let Fw be the event: First person chosen is a woman, Sw that second person chosen is a woman. Then

$$P\{Fw|Sw\} = P\frac{\{Fw \cap Sw\}}{P\{Sw\}} = \frac{P\{Sw|Fw\}P\{Fw\}}{P\{Sw\}}$$

$$= \frac{P\{Sw|Fw\}P\{Fw\}}{P\{Sw|Fw\}P\{Fw\} + P\{Sw|Fw^c\}P\{Fw^c\}}$$

$$= \frac{\dfrac{9}{24}\dfrac{10}{25}}{\dfrac{9}{24}\dfrac{10}{25} + \dfrac{10}{24}\dfrac{15}{25}} = \frac{9}{24}.$$

Note that this is the same as $P\{Sw|Fw\}$, which is not really surprising.

Px 4.23.1 Let F_1 be the event that the first experiment is a failure, and let $F_{(2)}$ be the event of exactly 2 failures. The desired probability is

$$P\{F_1|F_{(2)}\} = \frac{P\{F_1 \cap F_{(2)}\}}{P\{F_{(2)}\}}$$

Using the definition of statistical independence, and the rules of probability, we find

$$P\{F_{(2)}\} = \binom{3}{2}p(1-p)^2$$

the factors involving p representing the probability of any specific sequence with exactly 2 failures (and hence, one success) and the binomial coefficient being the number of sequences of length 3 with exactly 2 failures (whose value is 3, because such sequences are completely specified by the location of the one lonely success). On the other hand

$$P\{F_1 \cap F_{(2)}\} = P\{F_{(2)}|F_1\}P\{F_1\}.$$

But

$$P\{F_{(2)}|F_1\} = \binom{2}{1}p(1-p), P\{F_1\} = 1-p$$

using the same type of reasoning, this binomial coefficient being 2. So

$$P\{F_1|F_{(2)}\} = \frac{2(1-p)^2 p}{3p(1-p)^2} = \frac{2}{3}.$$

Does this result come as a surprise?

Px4.23.3 Using DeMorgans's law, Problem 4.5.1.d. on page 98, and the assumed statistical independence of the three events, we find

$$
\begin{aligned}
1 - P\{A \cup B \cup C\} &= P\{(A \cup B \cup C)^c\} \\
&= P(A^c \cap B^c \cap C^c) \\
&= P\{A^c\}P\{B^c\}P\{C^c\}
\end{aligned}
$$

Since the right hand side is greater than 0 and less than 1, it is impossible for the union of A, B, C to have probability 1 (or the left side of these equations would be 0, which it cannot be).

Px 4.27.1 3.5

Px 4.27.4 Let NG represent the net gain. Then
$E(NG) = (1000 - 1) \times 0.0005 + (-1) \times 0.9995 = -0.5$, i.e., a net loss of 50 cents — meaning that state lotteries are a rip-off of the bettor.

Px 4.28.1 $E(X) = 3.5, Var(X) = \frac{35}{12}, \sigma_X \cong 1.7078$

Px 4.28.4

$$Var(X) = E[(X-p)^2] = (1-p)^2 \times p + (0-p)^2 \times (1-p) = p(1-p)$$
$$\sigma_X = \sqrt{p(1-p)}$$

Px 4.29.2

    a. Since $\sigma_{\bar{X}} = \sigma_X/\sqrt{n}$ we need $n = 100$, so that $\sqrt{n} = 10$.

    b. We will take the easy route here. Since $0 \le p \le 1$, we have

$$\sigma_X = \sqrt{p(1-p)} \le 1$$

We want

$$P\{|\bar{X} - p| \le 0.01\} \ge 0.95 . \quad (*)$$

By the Chebychev inequality, Theorem 4.36 on page 133,

$$P\left\{|\bar{X} - p| \ge t\sigma_{\bar{X}}\right\} \le 1/t^2$$

or

$$P\left\{|\bar{X} - p| < t\sigma_{\bar{X}}\right\} \ge 1 - 1/t^2$$

But since $\sigma_{\bar{X}} = \sigma_X/\sqrt{n} = \dfrac{\sqrt{p(1-p)}}{\sqrt{n}}$ for the Bernoulli case, we see that

$$P\{|\bar{X} - p| < t/\sqrt{n}\} \ge 1 - 1/t^2 .$$

Matching this with inequality (*) we want to choose

$$1 - 1/t^2 = 0.95 \text{ and } t/\sqrt{n} = 0.01$$

or $t^2 = 20$ and $n = t^2/0.0001$, from which we find that $n = 200{,}000$.

This is much larger than it need be, as we can see by direct computation on the binomial distribution that $n \cong 9600$ will be adequate.

Px 4.32 In general we are assuming that $E(X_i) = 0.1$, $\sigma_{X_i} = 0.1$. From the Central Limit Theorem

$$P\left\{\sum_{i=1}^{60k} X_i > 20\right\} = P\left\{\sum_{i=1}^{60k} \frac{(X_i - 0.1)}{\sqrt{6k}} > \frac{20 - 6k}{\sqrt{6k}}\right\} \cong 1 - \Phi\left(\frac{20 - 6k}{\sqrt{6k}}\right)$$

where Φ is the standard normal distribution function, k represents the number of hours and

$$\sum_{i=1}^{60k} X_i$$

is the particle displacement at k hours after release.

For part a., $k = 1$, for b. $k = 2$, etc.

We will simplify the problem of part f. by finding m miles such that

$$1 - \Phi\left(\frac{m - 72}{\sqrt{72}}\right) \cong 0.01$$

i.e., that the first particle released has only a 1% chance of being further than 12 miles away. We will assume that the Central Limit Theorem gives accurate results (it doesn't always give accurate results) and that even if some other particles are ahead of the first one, it is not by much. Since $\Phi(2.326) = 0.99$, we set $(m - 72)/\sqrt{72} = 2.326$ which yields $m \cong 91.741$.

Px 4.32.3 Let X_i be the number of failures on trial i (0 or 1). We want to find

$P\left\{\sum\limits_{i=1}^{100} X_i > 15\right\}$. But by the Central Limit Theorem

$$P\left\{\sum_{i=1}^{100} X_i > 15\right\} = P\left\{\sum_{i=1}^{100} \frac{X_i - 0.2}{\sqrt{0.2 \times 0.8 \times 100}} > \frac{15 - 20}{4}\right\}$$

$$\cong 1 - \Phi(-1.25) \cong 0.8944$$

The exact (binomial) answer is 0.9196, which is reasonably close.

Px 4.32.5 If X_i is the number of good units produced on trial i (0 or 1), we want to choose n so that

$$P\left\{\sum_{i=1}^{n} X_i \geq 200\right\} \geq 0.95.$$

But by the Central Limit Theorem

$$P\left\{\sum_{i=1}^{n} X_i \geq 200\right\} = P\left\{\sum_{i=1}^{n} \frac{(X_i - 0.2)}{\sqrt{n \times 0.2 \times 0.8}} \geq \frac{200 - 0.2n}{\sqrt{n \times 0.2 \times 0.8}}\right\}$$

$$\cong 1 - \Phi\left(\frac{200 - 0.2n}{\sqrt{n \times 0.2 \times 0.8}}\right)$$

where Φ is the standard normal distribution function. Since $\Phi(-1.645) = 0.05$, it follows that we want to choose n so that

$$\frac{200 - 0.2n}{\sqrt{n \times 0.2 \times 0.8}} = -1.645,$$

or

$$0.2n - 1.645 \times 0.4 \times \sqrt{n} - 200 = 0$$

From the quadratic formula, it follows that n is one of the four integers adjacent to the two values

$$1.645 \pm \frac{\sqrt{(0.658)^2 + 160}}{0.4} = 1.645 \pm \frac{\sqrt{160.433}}{0.4}.$$

Since the negative value has no meaning we choose n as the smallest integer whose value is at least $(1.645 + 31.665)^2 = 33.31^2 = 1110$, i.e., $n = 1110$.

Px 4.39.2 Let X_i be the number of seconds gained by your watch on day i. You want to choose n so that

$$P\left\{ \sum_{i=1}^{n} X_i \geq 60 \right\} \cong 0.05$$

or

$$P\left\{ \sum_{i=1}^{n} X_i < 60 \right\} \cong 0.95.$$

From the Central Limit Theorem

$$P\left\{ \sum_{i=1}^{n} X_i < 60 \right\} = P\left\{ \sum_{i=1}^{n} \frac{(X_i - 1.5)}{\sqrt{0.1n}} < \frac{60 - 1.5n}{\sqrt{0.1n}} \right\} \cong \Phi\left(\frac{60 - 1.5n}{\sqrt{0.1n}} \right)$$

where Φ is the standard normal distribution function. Since $\Phi(1.645) = 0.95$, it follows that we want to choose $n > 0$ so that approximately $1.5n + 1.645\sqrt{0.1n} - 60 = 0$. From the quadratic formula, this yields $\sqrt{n} \cong 6.15$ or $n = 38$. That is, the Central Limit Theorem approximation says to reset your watch every 38 days, and you will rarely be as much as a minute *off* of the correct time.

Chapter 5

Px 5.1.1 Because $50 \times 0.9 = 45$ and $[45] = 45$ (see Definition 5.4 on page 147) using Equation 5.5 on page 147, we find that

$$P\{X_{[50 \times 0.9]} < q_{0.95}\} - 1 \quad P[S_{50} \leq 44\}$$

where S_{50} is binomial with $n = 50$ and $p = .95$. But from the binomial distribution (Calc -> Probability Distributions -> Binomial) we find

$$P\{S_{50} \leq 44\} = 0.0378.$$

So

$$P\{X_{[50 \times 0.9]} \leq q_{0.95}\} = 0.9622.$$

For $q_{0.975}$ from the binomial distribution with $n = 50$ and $p = .975$ we have

$$P\{X_{\underline{[50 \times 0.9]}} \leq q_{0.975}\} = 0.9985$$

Px 5.1.2 It appears that we can use such data to find high probability lower limits for these quantiles. That is $X_{[50 \times 0.9]}$ has a high probability of being less than $q_{0.975}$. This furnishes quite a bit of information about these quantiles (which might not be known to us without such data). These results are examples of *lower confidence limits*.

Prob 5.6.3 For $\lambda = 1, 10$ we put Poisson samples in columns $c1$ to $c100$. Then we generated the sample mean and sample variance of the rows in columns $c103$ and $c104$ respectively, by the commands shown.

```
SUBC> poisson 1.
MTB > rmean c1-c100 c103
MTB > rstdev c1-c100 c106
MTB > let c104 = c106*c106
MTB > describe c103-c104
```

Descriptive Statistics: C103, C104

Variable	N	Mean	Median	TrMean	StDev	SE Mean
C103	200	1.0027	1.0000	1.0019	0.1026	0.0073
C104	200	1.0108	0.9687	1.0033	0.1819	0.0129

Variable	Minimum	Maximum	Q1	Q3
C103	0.6800	1.3200	0.9400	1.0675
C104	0.6541	1.5332	0.8781	1.1352

```
MTB > random 200 c1-c100;
SUBC> poisson 10.
MTB > rmean c1-c100 c103
MTB > rstdev c1-c100 c106
MTB > let c104=c106*c106
MTB > describe c103-c104
```

Descriptive Statistics: C103, C104

Variable	N	Mean	Median	TrMean	StDev	SE Mean
C103	200	10.047	10.060	10.051	0.321	0.023
C104	200	10.075	10.025	10.044	1.447	0.102

The standard deviations of $c103$ and $c104$ (gotten by the describe command) seem to indicate that the sample mean is a better estimate than the sample variance. The density histograms below make this assertion even more evident. For $\lambda = 10$ the results are even more convincing.

The density histograms below are more reliable evidence for the above conclusions.

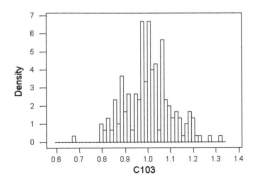

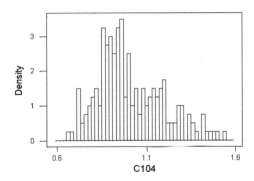

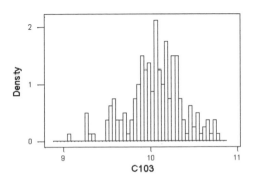

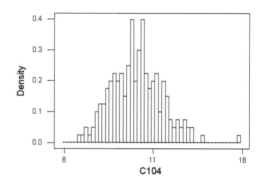

Px 5.4.3 If Z is a standard normal variable, we can see from the standard normal distribution function (Calc -> Probability Distributions -> Normal ...) that

$$P\{|Z| \le 2.5758\} = 0.99.$$

$$P\{|\bar{X}_n - \mu| \le k\} = P\left\{\left|\frac{\bar{X}_n - \mu}{15/\sqrt{n}}\right| \le \frac{k\sqrt{n}}{15}\right\}$$

We chose $k = 5$, and by Equation 4.34 on page 136, we see that we want

$$\frac{k\sqrt{n}}{15} = 2.5758$$

from which we find $n = 46$.

Prob 5.11.1 The $1 - \alpha$ level confidence interval for the mean μ is

$$\bar{X}_n \pm t_{n-1, 1-\alpha/2} s_n / \sqrt{n}$$

where $t_{n-1, 1-\alpha/2}$ is the upper $1 - \alpha/2$ point of the student t distribution with $n-1$ degrees of freedom (see Definition 5.9 on page 155)

Prob 5.11.2 The $1 - \alpha$ level confidence interval for the standard deviation ratio σ_x / σ_y is

$$\left(\frac{s_{x,n}}{s_{y,m}} \frac{1}{\sqrt{F_{n-1, m-1, 1-\alpha/2}}}, \frac{s_{x,n}}{s_{y,m}} \frac{1}{\sqrt{F_{n-1, m-1, \alpha/2}}}\right)$$

where for each value β, and all integers n,m exceeding $1, F_{n-1, m-1, 1-\beta}$ is the upper $1 - \beta$ point ((see Definition 5.9 on page 155) of the F distribution with $n-1$ and $m-1$ degrees of freedom. The quantities $s_{x,n}$ and $s_{y,m}$ are the sample standard deviations of the x and y samples of size n and m respectively. These intervals can be used to decide whether or not two normal random variables have the same standard deviations (which arises in checking the usual ANOVA assumptions).

Prob 5.11.3 The $1 - \alpha$ level confidence interval for the variance σ^2 of a normal distribution is

$$\left((n-1)s_n^2 \frac{1}{\chi_{n-1,\,1-\alpha/2}^2}, \ (n-1)s_n^2 \frac{1}{\chi_{n-1,\,\alpha/2}^2} \right)$$

where for each value β, and all integers n exceeding $1, \chi_{n-1,\,1-\beta}^2$ is the upper $1 - \beta$ point (see Definition 5.9 on page 155) of the *Chi-Square* distribution with $n - 1$ degrees of freedom.

Px 5.6.2 We find out which intervals of Problem 5.11.3 include the value $\sigma^2 = 36$ as follows. Assuming that column c21 consists of the lower confidence limits for σ^2 and column c23 the upper confidence limits for σ^2, there are three cases

	sign of 36 - c21 value	sign of c23 value - 36
$\sigma^2 = 36$ between limits	+	+
$\sigma^2 = 36 <$ lower limit	-	+
$\sigma^2 = 36 >$ upper limit	+	-

elements which are positive correspond to being within the limits. Here is the complete code for the case actually run.

```
MTB > random 100 c1-c15;
SUBC> normal 3 6.
MTB > rstdev c1-c15 c18
MTB > let c19=c18*c18
MTB > let c21 = c19*.591
MTB > let c23 = c19*2.131
MTB > let c25 = 36 -c21
MTB > let c26 = c23 -36
MTB >
MTB > let c28 = c25*c26
MTB > let c29 = ((c28/abs(c28))+1)/2
MTB > mean c29
```

Mean of C29

```
Mean of C29 = 0.92000
```

In this case 92% of the values were within the confidence limits.

Px 5.6.3

a. $n = 10$, $\chi_{9,\,0.995}^2 = 23.5894$, $\chi_{9,\,0.005}^2 = 1.7349$, $t_{9,\,0.995} = 3.2498$

(The standard normal upper .995 point is 2.57...)

A reasonable assumption might be that the data is normal, mean μ, standard deviation σ, both unknown. True, the histogram looks somewhat uniform, but there is not enough data to draw firm conclusions, and normal theory often holds up well in practice. $\bar{X}_{10} = 17.18$, $s_{10} = 1.524$.

Using the Minitab one-sample t test,

the 99% confidence interval for μ is   [15.614, 18.746]

and the 99% confidence interval for σ is   [.94, 3.47].

b. You can use the upper value of the confidence interval for σ and the lower value of the confidence limit for μ (these being worst with respect to mileage) remembering that these were .995 level limits. Then you are at least 98.5% sure that your actual mileage on any one tank (or, as we ran it, on any 3/4 tank) will be at least $15.614 - 2.5758 \times 3.47 = 7.6579$ mpg. We assume that the car has a 20 gallon tank, but we're only relying on 15 gallons (allowing 5 gallons to reach a gas station). The total miles that we can be 98.5% sure of getting safely is $7.6579 \times 15 = 115$ miles. Here we used Bonferroni's inequalities (Theorem 4.8 on page 103) to get the number 98.5% (with the 2.5758 coming from the standard normal tables).

Px 5.7.2

```
MTB > random 100 c1-c201;
SUBC> unif -.5 .5.
MTB > rmean c1-c201 c204
MTB > rmedian c1-c201 c205
MTB > describe c204-c205
```

Descriptive Statistics: C204, C205

Variable	N	Mean	Median	TrMean	StDev	SE Mean
C204	100	-0.00071	-0.00149	-0.00056	0.02323	0.00232
C205	100	-0.00189	-0.00741	-0.00143	0.03724	0.00372

Variable	Minimum	Maximum	Q1	Q3
C204	-0.06242	0.05911	-0.01465	0.01650
C205	-0.11108	0.08042	-0.02415	0.02320

From this output, and the reasoning mentioned in the question itself, it seems pretty evident that the mean is better than the median when the distribution is uniform.

Px 5.7.3 With the t distribution having 3 degrees of freedom it is evident, from the table to follow, that the sample median is a far better estimate of the mean than is the sample mean.

```
MTB > random 100 c1-c201;
SUBC> t 3.
MTB > rmean c1-c201 c203
MTB > rmed c1-c201 c204
MTB > describe c203-c204
```

Descriptive Statistics: C203, C204

Variable	N	Mean	Median	TrMean	StDev	SE Mean
C203	100	-0.0065	0.0045	-0.0081	0.1278	0.0128
C204	100	-0.0072	-0.0028	-0.0061	0.1000	0.0100

Variable	Minimum	Maximum	Q1	Q3
C203	-0.2687	0.4071	-0.1101	0.0785
C204	-0.2621	0.2042	-0.0795	0.0523

Px 5.9.1 If the people my brother-in-law met were a random sample from the population of those who make their own wine, then his claim translates to *the probability of a person from this population wanting to purchase his winemaker may be as high as .1* (implying that it isn't likely to be much higher). To see whether he was in contact with reality, we construct a .95 level confidence interval for *p* based on the 18 out of 20 people expressing a desire to purchase. The value of the .95 level confidence interval for *p* is [.683, .988]. It indicates that with high probability the demand is almost 7 times as much as my brother-in-law thought. Such a situation might antagonize the potential buyers if they are unable to purchase his device — which will be the case if he gears up for only 1/7 of the actual demand. Such miscalculation can ruin a business.

Px 5.9.4 If the dice were fair, the probability of an even sum is .5. The .95 level confidence interval for the probability of an even sum, based on the 130 even sums in 200 tosses is [.5805, .7159]. I'd wager good money that the dice are loaded.

Px 5.12.1.b The essential ideas behind our use of histograms to compare estimators are fairly simple — although the details necessary to nail down a completely trustworthy procedure would (if we had the time) require some pretty advanced mathematics. So we'll cut corners and present the heart of the reasoning, followed by an outline of what details need to be filled in. If we had extremely large samples of both estimators, under reasonable conditions the histograms would be good approximations to the densities of these estimators. So we act as if we really had available the estimators' densities. These densities would give us all the information we need to judge the estimators. For instance, from each density we could compute the probability that the given estimator is within any specified distance from the quantity being estimated. (Of course we must do all such estimations one at a time.) Alternatively if we specified a probability, we could find a symmetric interval about the estimator having that probability. Presumably the smaller interval would yield (for that probability) the better estimator. Of course it may turn out that one estimator is preferable for one specified probability, and the other estimator is preferable for some other specified probability. Or for a specified length interval about the estimator, the probability is greater for one of the estimators, and for a different specified length it is greater for the other. And all this is for one specified value of the unknown parameter. Matters might be altered completely for another one of the parameter's values. All of this simulation wouldn't be unnecessary if we could analytically compute the distributions of the estimators for all parameters' values. But we usually cannot do this, and simulation comes close to doing this job, because, at the cost of big chunks of computer time, adequate (if not perfect) accuracy can usually be achieved.

Since this was only a demonstration, and there was no special reason to overtire our computer, we chose a relatively small number of observations (51) and simulations (1600). Nonetheless, the results are quite convincing.

We will only present the results of comparing the mean and the median for the t distributions with 2 and 12 degrees of freedom.

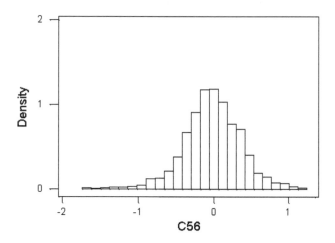

Histogram for Mean of t with 2 d.f.

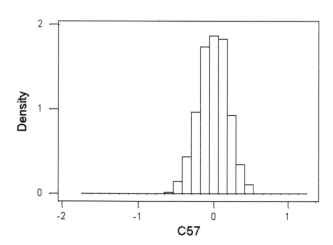

Histogram for Median of t with 2 d.f.

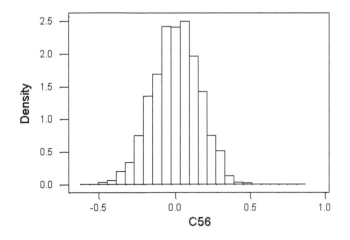

Histogram of Mean of t with 12 d.f.

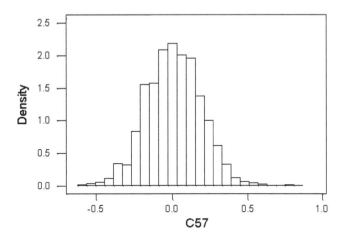

Histogram of Median of t with 12 d.f.

It looks clear that for 2 degrees of freedom, the sample median wins hands down. It is a little harder to see, but the sample mean appears better in the case of 12 degrees of freedom.

Nailing down the details is done with the help of confidence bands for the cumulative distribution function, based on the Kolmogorov-Smirnov theorem, which utilizes the empirical cumulative distribution function. With sufficiently many simulations, this allows arbitrarily accurate estimation of a distribution function from the empirical cumulative distribution function. From this one can determine interval probabilities with as much accuracy as is desired.

Px 5.12.2 Using the histograms shown below, it would appear that the average of the maximum and minimum beats the mean to estimate the population mean when the distribution is uniform (see answer to previous exercise).

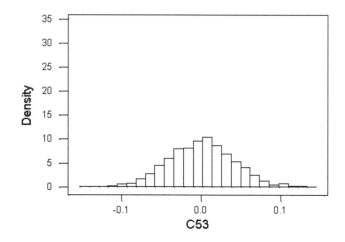

Histogram of Uniform Distribution Means

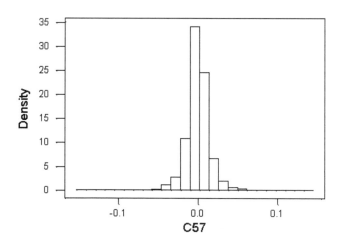

Histogram of (Max + Min)/2 for Uniform Distribution

Chapter 6

Px 6.3.1 The class would seem to consist of all location shifts of some specified distribution. Usually just one of these distributions is the null hypothesis. A more reasonable choice would be the set of distributions shifted by less than or equal to some specified value, v, from the one specified in the previous null hypothesis.

Px 6.4.2

a. The test is to reject H_0 if $\left|\dfrac{\overline{X}-0}{3/\sqrt{5}}\right| > 1.96$.

   In the given case, the X_i were .15479, -3.5498, 3.79334, 3.30389, .62212, so $\overline{X} = 0.86487$ and we do not reject H_0.

b. Here the test is to reject H_0 if, $\dfrac{\overline{X}-0}{3/\sqrt{5}} < 1.645$ and we do not reject H_0 here either.

Px 6.5.2 We give the commands of the first session window of Topic 6.2 with 10 replaced by 7. The result is as shown.

Paired T-Test and CI: C1, C2

Paired T for C1 - C2

	N	Mean	StDev	SE Mean
C1	7	172.66	9.51	3.60
C2	7	160.09	14.21	5.37
Difference	7	12.57	5.82	2.20

95% CI for mean difference: (7.19, 17.95)
T-Test of mean difference = 0 (vs not = 0): T-Value = 5.72 P-Value = 0.001

So the null hypothesis (of *no effect*) is still rejected at any significance level above .001.

Px 6.5.4 For the case $\sigma = 10$ the *p*-value is .000, so there is still no rejection of the null hypothesis (of *no effect*) at any significance level above .001. When $\sigma = 30$ the confidence interval for the mean of the difference is (.01, 38.64), so there is still no rejection at the 5% significance level; but this will vary because of sampling error having a big effect in such a borderline case.

Px 6.6.1 The data was generated in exactly the same way as done in the answer to Px 6.5.2., but the differences had to be taken by us before we could use the one-sample Wilcoxon test (the paired *t* computing the differences itself). The results are as shown below, which indicate that at the usual significance level of .05 the null hypothesis of *no effect* would still be rejected, but with less so-called significance. Typically under the normality assumption, this Wilcoxon test is somewhat less sensitive to changes in the mean than the one-sample *t* test.

```
Wilcoxon Signed Rank Test: C4


Test of median = 0.000000 versus median not = 0.000000

                      N for    Wilcoxon                 Estimated
                N    Test   Statistic          P         Median
     C4         7     7         0.0         0.022        -12.34
```

Px 6.8.1

```
MTB > random 6 c1-c5;
SUBC> normal 35 8.
MTB > random 9 c6-c10;
SUBC> normal 41 8.
we test c1 vs c6, c2 vs c7  etc
|
MTB > TwoSample c1 c6;
SUBC>    Pooled.
95% CI for difference: (-15.35, -1.63)
T-Test of difference = 0 (vs not =): T-Value = -2.67  P-Value = 0.019  DF = 13

MTB > twosample c2 c7;
SUBC> pooled.
95% CI for difference: (-14.01, 1.72)
T-Test of difference = 0 (vs not =): T-Value = -1.69  P-Value = 0.115  DF = 13

MTB > twosample c3 c8;
SUBC> pooled.
95% CI for difference: (-18.07, -7.54)
T-Test of difference = 0 (vs not =): T-Value = -5.25  P-Value = 0.000  DF = 13
```

Notice that in cases like this, where the null hypothesis is not true, the actual state of affairs can be borderline, with the result that you can be wrong about half the time.

Px 6.9.2 We will just do part b because the confidence intervals will tell whether the null hypothesis was rejected at the .05 significance level — rejection at that level occurs if the interval does not include the value 0. The actual inputs and outputs are presented next, followed by some comments on what it all seems to mean.

```
MTB > random 6 c1;
SUBC> normal 35 4.
MTB > random 12 c2;
SUBC> normal 51 14.
MTB > random 70 c3;
SUBC> normal 35 4.
MTB > random 30 c4;
SUBC> normal 51 14.
```

Having generated the data, we perform the tests, with the results shown.

```
MTB > twosample c1 c2;
SUBC> pooled.
```

95% CI for difference: (-34.56, -2.08)
```
T-Test of difference = 0 (vs not =): T-Value = -2.39  P-Value = 0.029  DF = 16
Both use Pooled StDev = 15.3
```

```
MTB > twosample c1 c2
```

Estimate for difference: -18.32
```
95% CI for difference: (-30.32, -6.32)
T-Test of difference = 0 (vs not =): T-Value = -3.30  P-Value = 0.006  DF = 13
```

```
MTB >
MTB > twosample c3 c4
95% CI for difference: (-23.47, -13.39)
T-Test of difference = 0 (vs not =): T-Value = -7.47  P-Value = 0.000  DF = 30
```

```
MTB > twosample c3 c4;
SUBC> pooled.
```

95% CI for difference: (-21.83, -15.02)
```
T-Test of difference = 0 (vs not =): T-Value = -10.74  P-Value = 0.000  DF = 98
Both use Pooled StDev = 7.86
```

```
Both use Pooled StDev = 6.46
MTB > twosample c2 c3
95% CI for difference: (7.88, 17.55)
T-Test of difference = 0 (vs not =): T-Value = 5.64  P-Value = 0.000  DF = 14
MTB > twosample c1 c4;
SUBC> pooled.
95% CI for difference: (-16.772, -13.819)
T-Test of difference = 0 (vs not =): T-Value = -20.83  P-Value = 0.000  DF = 48
Both use Pooled StDev = 2.08
MTB > twosample c1 c4
95% CI for difference: (-16.287, -14.304)
T-Test of difference = 0 (vs not =): T-Value = -31.47  P-Value = 0.000  DF = 31
MTB > twosample c2 c4;
SUBC> pooled.
95% CI for difference: (-1.76, 2.56)
T-Test of difference = 0 (vs not =): T-Value = 0.37  P-Value = 0.715  DF = 78
MTB > twosample c2 c4
95% CI for difference: (-1.78, 2.58)
T-Test of difference = 0 (vs not =): T-Value = 0.37  P-Value = 0.715  DF = 48
MTB > twosample c3 c4;
SUBC> pooled
95% CI for difference: (-14.75, -9.89)
T-Test of difference = 0 (vs not =): T-Value = -10.20  P-Value = 0.000  DF = 48
Both use Pooled StDev = 3.42
MTB > twosample c3 c
95% CI for difference: (-16.93, -7.70)
T-Test of difference = 0 (vs not =): T-Value = -6.04  P-Value = 0.000  DF = 9
MTB >
```

px6.8.1.mtw ***

+	C1	C2	C3	C4	C5	C6	C7
1	33.8399	58.0963	32.3471	47.7369			

We extract the following results from the data here.

columns being tested	pooled variance confidence interval	unpooled variance confidence interval
c1 c2	(-19.6, -11.53)	(-17.86, -13.52)
c1 c3	(-7.25, 1.30)	(-7.58, 1.63)
c1 c4	(-16.772, -13.819)	(-16.287, -14.304)
c2 c3	(8.13, 17.31)	(7.88, 17.55)
c2 c4	(-1.76, 2.56)	(-1.78, 2.58)
c3 c4	(-14.75, -9.89)	(-16.93, -7.7)

We see that in the cases considered, the two tests seem to give very similar results, even when the standard deviations of the two groups are quite different. A comparison of these tests that will probably make it easier to decide which to use, is to compute their power at alternatives of interest (see Section 6.4 starting on page 209).

Prob 6.8 Even if the distributions are quite different, the independent two-sample t test may have very little chance of detecting such a difference. For this t test to detect a difference, there must be some location shift that is somewhat greater than the standard deviation of the numerator of the t statistic. To see this, just let F_X be a normal distribution with mean 0 and standard deviation 1, and let F_Y be normal with mean 0 and standard deviation 50. The independent two-sample t test is blind to such a difference.

Prob 6.10 The distributions in the answer to Problem 6.8 above show that the two-sample Wilcoxon test also does not discriminate against distributions that are not shifted in location from each other.

Px 6.13.2 In the macro *iwilpow.mac* delete the semicolon on the line preceding the one that starts   cauchy 0 3, and delete the line that follows that semicolon. Replace the line   *cauchy k10 k11.*   with *exponential 3.*
Before running the program, in the session window type

    let k1 = 13

    let k2 = 17

    let k3 = 1000

    let k8 = 1154.67

    let k9 = 248.33

Our power estimate against the exponential 3 distribution was .985 based on 400 simulation trials. From this we obtain the 95% level confidence interval (.968,.994), using the binomial confidence interval construction.

To obtain the actual significance level (as opposed to the nominal level which is calculated from a Central Limit Theorem normal distribution approximation) yields the estimate .06. The associated 95% confidence interval is (.039, .088).

Both confidence intervals could be made as small as desired, with confidence as large as desired (excluding, of course, 0 length and 100% confidence) by running sufficiently many trials in the Monte Carlo simulations. We used 400 trials here.

Chapter 7

Px 7.2.2 The data for this exercise was generated as shown below.

```
MTB > set cl
DATA> 45:75/3
DATA> end
MTB > let c2 = cl*cl
MTB > let c3 = -15 + 2*cl -.02222*c2
MTB > random 11 c4;
MTB > normal 0 2.
MTB > let c5 = c3+c4
MTB > plot c5*cl c3*cl;
SUBC> symbol;
SUBC> type 1 0;
SUBC> line cl c3;
SUBC> overlay.
```

Worksheet 1 ***

	C1	C2	C3	C4	C5	C6	C7	C8
1	45	2025	30.0045	-1.30445	28.7000			
2	48	2304	29.8051	0.83000	30.6351			
3	51	2601	29.2058	1.74197	30.9477			
4	54	2916	28.2065	3.42596	31.6324			
5	57	3249	26.8072	1.78122	28.5884			
6	60	3600	25.0080	-3.31491	21.6931			
7	63	3969	22.8088	-0.86160	21.9472			
8	66	4356	20.2097	-1.58701	18.6227			
9	69	4761	17.2106	0.94366	18.1542			
10	72	5184	13.8115	0.65314	14.4647			
11	75	5625	10.0125	-1.31678	8.6957			

The first coordinates of the *raw data* are in column c1, the second coordinates of the raw data are in column c5, while the second coordinates of the underlying data (without the noise which was added to get the *genuine* raw data) are in column c3. We then plotted both c5 vs c1 and c3 vs c1. It is worth mentioning the philosophy behind the line commands (which I find far simpler than the *point and click* way of generating overlays). In the first line you put each of the set of pairs you want to plot, for example   ck versus cj   in the form   ck*cj (separated from other pairs by spaces).

To specify what symbol you want each set of pairs to be plotted with, use the *symbol* command and its *type* subcommand i.e.,

    symbol;

    type 1 0.

The *type* subcommand affects of ordered pair sets in the order given in the plot command, so that the value 1 affects c5 versus c1, while the number 0 affects c3 versus c1. The 1 indicates use of the first symbol (to be used on the raw data). The 0 indicates that no symbol is to be used for the true underlying relation. Then line c1 c3 indicates connecting the points from the true relation. We refrain from showing this plot because next we do the regression, which uses the raw data to obtain a fitted second degree polynomial. We will show that plot which Minitab generates, and then we will plot the genuine raw data, the curve from which it was generated and the fitted curve. The regression is invoked from the menu bar via

    Stat -> Regression -> Fitted Line Plot

which led to the dialogue box that we filled out as indicated, and pressed OK.

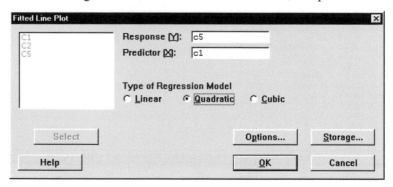

This yielded the following output.

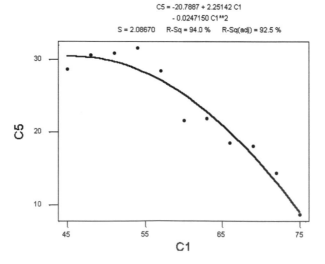

Now that we had the fitted regression equation,
$$c5 = -20.7887 + 2.25142\ c1 - .0247150\ c1*c1$$
we added this curve to the plot. To do this, notice that we already had column $c1$ filled with the first (x) coordinates of everything being plotted; column $c3$ was already filled with the second (y) coordinates of the true relation, and column $c5$ was filled with the actual data second coordinates. So we had to use column $c6$ for the second coordinates of the fitted equation ($c5$ already being used). Here is what we typed to get the plot to follow (to extend the way to plot overlays).

plot c1*c3 c1*c5 c1*c6;

symbol;

type 0 1 0;

connect;

type 1 0 2;   (1 stands for solid line connection the c1*c3 points (first listed plot)
              0 stands for no connection of the c1*c3 points (second listed plot),
              and 2 is for dashed connection of the c1*c6 points (third listed plot)).

overlay.

Here is what resulted.

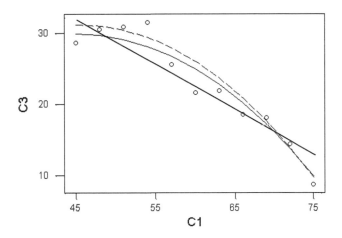

Notice how close the fitted curve (dashed) is to the true curve. Just fitting a straight line by eye shows that no straight line can be nearly as close to the true curve (solid) as the fitted curve (dashed one) is.

Px 7.2.3 Just keep the same relationships as in the previous exercise, but keep raising the standard deviation of the random error added, until the fitted curve gets pretty far from the true one.

Px 7.5.1 The output was the same as if the additive two-way layout was chosen. The reason that the given design was not treated as a general two-way layout was outlined in the discussion of Topic 7.4 — namely, too small a sample size.

Px 7.5.2 Below is the data with the information about how it was created.

↓	C1 measured bp	C2 patient num	C3 dose level	C4	C5 patient effect	C6 avg treatment dose effect	C7 patient - dose interaction	C8 random error
1	150.244	1	1		150	0	0	0.24393
2	152.419	1	2		150	10	0	-7.58068
3	160.539	1	3		150	13	0	-2.46141
4	158.178	1	1		150	0	0	8.17770
5	166.181	1	2		150	10	0	6.18114
6	163.156	1	3		150	13	0	0.15630
7	154.844	2	1		160	0	0	-5.15610
8	177.843	2	2		160	10	9	-1.15698
9	199.114	2	3		160	13	19	7.11404
10	162.012	2	1		160	0	0	2.01200
11	179.413	2	2		160	10	9	0.41347
12	190.390	2	3		160	13	19	-1.61029
13	173.320	3	1		170	0	0	3.31966
14	180.421	3	2		170	10	0	0.42099
15	184.427	3	3		170	13	0	1.42727
16	165.956	3	1		170	0	0	-4.04433
17	178.475	3	2		170	10	0	-1.52505
18	186.863	3	3		170	13	0	3.86285
19	161.640	4	1		165	0	0	-3.35972
20	169.007	4	2		165	10	0	-5.99335
21	181.281	4	3		165	13	0	3.28064
22	165.672	4	1		165	0	0	0.67198
23	170.587	4	2		165	10	0	-4.41287
24	177.701	4	3		165	13	0	-0.29876

Following the creation of the data is the command to run the two-way analysis of variance, followed by the Minitab output.

```
MTB > let c1 = c5+c6+c7+c8
MTB > twoway c1 c2 c3
```

Two-way ANOVA: measured bp versus patient num, dose level

```
Analysis of Variance for measured
Source       DF      SS       MS       F       P
patient       3    1493.9    498.0   24.66   0.000
dose lev      2    1440.2    720.1   35.66   0.000
Interaction   6     460.7     76.8    3.80   0.023
Error        12     242.3     20.2
Total        23    3637.2
```

The test that all interactions are 0 has p-value .023, indicating that there are significant interactions — i.e., that the test concludes that at least one interaction is not 0. Minitab however does not give any indication of the interactions sizes. There are several reasons for this. First, it's a tough problem, but sometimes solvable in a satisfactory way, with a lot of effort. Second, in most cases the hope is that all interactions are 0, so as to simplify the model for later use.

Index

T - #0398 - 071024 - C292 - 234/156/13 - PB - 9780367396459 - Gloss Lamination